The Open University

M3?

B2

Subdivisions

This publication forms part of an Open University course. Details of this and
other Open University courses can be obtained from the Student Registration
and Enquiry Service, The Open University, PO Box 197, Milton Keynes,
MK7 6BJ, United Kingdom: tel. +44 (0)870 333 4340, e-mail
general-enquiries@open.ac.uk

Alternatively, you may visit the Open University website at
http://www.open.ac.uk where you can learn more about the wide range of
courses and packs offered at all levels by The Open University.

To purchase a selection of Open University course materials, visit the webshop
at www.ouw.co.uk, or contact Open University Worldwide, Michael Young
Building, Walton Hall, Milton Keynes, MK7 6AA, United Kingdom, for a
brochure: tel. +44 (0)1908 858785, fax +44 (0)1908 858787, e-mail
ouwenq@open.ac.uk

The Open University, Walton Hall, Milton Keynes, MK7 6AA.

First published 2006.

Copyright © 2006 The Open University

All rights reserved; no part of this publication may be reproduced, stored in a
retrieval system, transmitted or utilised in any form or by any means, electronic,
mechanical, photocopying, recording or otherwise, without written permission from
the publisher or a licence from the Copyright Licensing Agency Ltd. Details of such
licences (for reprographic reproduction) may be obtained from the Copyright
Licensing Agency Ltd, 90 Tottenham Court Road, London W1T 4LP.

Open University course materials may also be made available in electronic formats
for use by students of the University. All rights, including copyright and related
rights and database rights, in electronic course materials and their contents are
owned by or licensed to The Open University, or otherwise used by The Open
University as permitted by applicable law.

In using electronic course materials and their contents you agree that your use will
be solely for the purposes of following an Open University course of study or
otherwise as licensed by The Open University or its assigns.

Except as permitted above you undertake not to copy, store in any medium
(including electronic storage or use in a website), distribute, transmit or re-transmit,
broadcast, modify or show in public such electronic materials in whole or in part
without the prior written consent of The Open University or in accordance with the
Copyright, Designs and Patents Act 1988.

Edited, designed and typeset by The Open University, using the Open University
T_EX System.

Printed and bound in the United Kingdom by The Charlesworth Group,
Wakefield.

ISBN 0 7492 4130 6

1.1

Contents

Introduction	**4**
Study guide	5
1 Regular subdivisions	**6**
1.1 The regular subdivision formulas	7
1.2 Dual subdivisions	10
1.3 Regular subdivisions of the sphere	12
1.4 Regular subdivisions of the torus	15
1.5 Surfaces with negative Euler characteristic	16
2 Finding the characteristic numbers	**23**
2.1 Finding the Euler characteristic	23
2.2 Finding the boundary number	25
2.3 Finding the orientability number	25
3 Edge expressions	**32**
3.1 The idea of an edge expression	32
3.2 The characteristic numbers	35
3.3 Edge equations	37
4 Strict subdivisions	**41**
4.1 Triangulations	41
4.2 Invariance of the Euler characteristic	47
Solutions to problems	**50**
Index	**55**

Introduction

In *Unit B1, Surfaces*, we stated the Classification Theorem for compact surfaces, which asserts that a surface is classified by its three characteristic numbers: the *boundary number* β, the *orientability number* ω and the *Euler characteristic* χ. Our aim here is to prepare the ground for a proof of the Classification Theorem, which we present in the next unit.

We explain how to determine the characteristic numbers of a surface given as a polygon with edge identifications. Then we show how a geometric description of a surface as a polygon with edge identifications can be replaced by an algebraic description. Although this algebraic description is little more than a list of the edges in order as we go around the boundary of the polygon, it turns out to be a surprisingly effective way of handling homeomorphisms of surfaces, as you will see in the next unit, and it plays a crucial role in the proof of the Classification Theorem.

In *Unit B1* we introduced the idea of a *subdivision* of a surface. By counting the number of vertices (V), edges (E) and faces (F) in a subdivision, we can find the Euler characteristic $V - E + F$ of the subdivision. We claimed that any two subdivisions of the same compact surface have the same Euler characteristic: thus, the Euler characteristic is a property of the surface, and not just of the subdivision. We stated this without proof in *Unit B1*, and return to it in this unit.

Before we start on the agenda outlined above, we consider a special type of subdivision — a *regular subdivision* of a surface without boundary. You may already be familiar with the regular subdivisions of a sphere: the tetrahedron, cube, octahedron, dodecahedron and icosahedron. For a regular subdivision, the faces are all polygons of the same type and the vertices have the same number of edges emerging from them. We can also show that any surface without boundary admits a *finite number* of regular subdivisions: this is a striking generalization of the result for the sphere. This result then allows us to answer such questions as: does there exist a regular hexagonal subdivision with exactly four hexagons meeting at each vertex on a surface without boundary and with Euler characteristic -4?

Study guide

In Section 1, *Regular subdivisions*, we consider regular subdivisions of a surface without boundary. This section is mainly computational, and you should make sure that you understand the methods involved.

The classification of a surface depends on its characteristic numbers, and in Section 2, *Finding the characteristic numbers*, we discuss techniques for finding the Euler characteristic and boundary number of a surface. We also investigate further what it means for a surface to be orientable, describing techniques for finding Möbius bands in surfaces. You should make sure that you understand all these techniques.

In Section 3, *Edge expressions*, we move from presenting a surface geometrically as a polygon with edge identifications to describing it in algebraic terms. We do this by writing down the edges in the order in which they occur, taking note of their directions. This description is called an *edge expression* for the surface, and is simple but surprisingly useful. We show you how to find the characteristic numbers of a surface from an edge expression. You should make sure that you understand how to deal with edge expressions, since you will meet them again in the next unit.

Finally, Section 4, *Strict subdivisions*, builds on our work in *Unit B1* on subdivisions and is more theoretical in nature. It is mainly a reading section, and introduces the idea of a *triangulation* of a strict subdivision. The use of triangulations helps us to understand why the Euler characteristic of a surface is independent of the subdivision chosen. If you are short of time, you may prefer to skim through this section on a first reading, and return to it when you need to.

There is no software associated with this unit.

1 Regular subdivisions

After working through this section, you should be able to:
▶ explain what is meant by a *regular subdivision*;
▶ derive the regular subdivision formulas for a given surface without boundary;
▶ construct the *dual* of a given subdivision;
▶ find all the regular subdivisions of a given kind on a given surface without boundary.

In *Unit B1* you met the idea of a *subdivision* of a surface, which we defined as follows.

Definition

A **subdivision** of a surface consists of a finite set of **vertices** and a finite set of **edges** such that:
- each vertex is an endpoint of at least one edge;
- each endpoint of an edge is a vertex;
- the vertices and edges form a connected graph;
- no two edges have any points in common;
- if the surface has a boundary, the boundary consists only of vertices and edges;
- the space obtained from the surface by removing the vertices and edges is a union of a finite number of disjoint pieces, called **faces**, each of which is homeomorphic to an open disc.

For example, Figure 1.1 shows a subdivision of the sphere, with eight vertices, twelve edges and six faces. This subdivision has a certain amount of regularity: all the faces have the same number of edges (they are all four-sided), and all the vertices look the same (each vertex has three faces meeting there, and three edges emerging from it). In this section we consider subdivisions, with similar regularity properties, of the sphere, the torus and other surfaces without boundary. We also consider the dual of a subdivision.

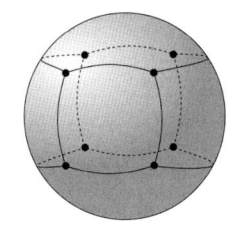

Figure 1.1

1.1 The regular subdivision formulas

We first define a *regular* subdivision.

> **Definition**
>
> A **regular subdivision** of a surface without boundary is a subdivision in which, for $j, k \geq 2$:
> - each face has the same number k of edges;
> - each vertex has the same number j of incident edges.

Recall that incident edges are those that meet at the vertex.

For example, the subdivision of the sphere in Figure 1.1 is a regular subdivision with $k = 4$ and $j = 3$.

The classical examples of regular subdivisions are those of the sphere provided by the regular polyhedra (see Figure 1.2):

- *tetrahedron*   4 triangular faces ($k = 3$),
  with 3 edges at each vertex ($j = 3$);
- *cube*   6 square faces ($k = 4$),
  with 3 edges at each vertex ($j = 3$);
- *octahedron*   8 triangular faces ($k = 3$),
  with 4 edges at each vertex ($j = 4$);
- *dodecahedron*   12 pentagonal faces ($k = 5$),
  with 3 edges at each vertex ($j = 3$);
- *icosahedron*   20 triangular faces ($k = 3$),
  with 5 edges at each vertex ($j = 5$).

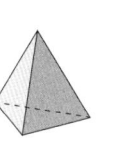

Later in this section we shall see that these are the only regular subdivisions of the sphere, except for two 'degenerate families'.

tetrahedron   cube   octahedron   dodecahedron   icosahedron

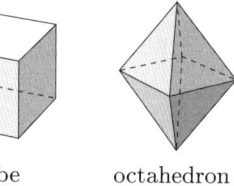

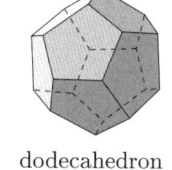

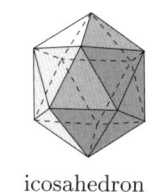

Figure 1.2   *These polyhedra are all topologically equivalent to the sphere.*

In order to prove results on regular subdivisions, we need to specify how we count the edges at a vertex in a given subdivision. What we do is to look at a small neighbourhood of the vertex, containing no other vertex, and count the edges that we see. If an edge starts and finishes at the same vertex, then we count this edge twice — we can think of counting 'edge ends'. For example, in Figure 1.3 the number of edges at the vertex P is 5.

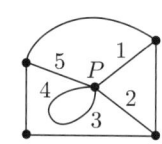

Figure 1.3

We can similarly count the edges around a face in a given subdivision. Since the face may meet itself along an edge, we go around the face, starting at an arbitrary vertex, until we return to the starting vertex. This may involve travelling along an edge twice: for example, the face F in Figure 1.4 is surrounded by 9 edges, because one edge is counted twice.

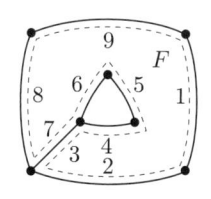

Figure 1.4

Another example is given by the subdivision of the torus in Figure 1.5. This subdivision has two vertices (P, Q), four edges (e_1, e_2, e_3, e_4), and two faces (F_1, F_2). Small neighbourhoods of the vertices P and Q are shown on the right: each vertex has four incident edges, one of them being counted twice. Starting near P and going along close to edge e_2, we see that the face F_1 is surrounded by the four edges e_2, e_3, e_2, e_1 (the edge e_2 being counted twice). Similarly, starting near Q and going along close to edge e_4, we see that the face F_2 is surrounded by the four edges e_4, e_1, e_4, e_3.

Note that this is a regular subdivision of the torus, because each face is surrounded by the same number of edges ($k = 4$) and each vertex has the same number of incident edges ($j = 4$).

Figure 1.5

The main result of this subsection is a collection of simple formulas that relate the Euler characteristic of a surface with a regular subdivision to the numbers k and j. These formulas give *necessary* conditions for a surface to admit a regular subdivision.

Theorem 1.1 The regular subdivision formulas

Let S be a surface without boundary with Euler characteristic χ, and having a regular subdivision with V vertices, E edges and F faces, in which each face has k edges and each vertex has j incident edges, where $j, k \geq 2$. Then $E = \frac{1}{2}Fk = \frac{1}{2}Vj$ and

$$\chi = E\left(\frac{2}{j} + \frac{2}{k} - 1\right) = \frac{1}{2}Fk\left(\frac{2}{j} + \frac{2}{k} - 1\right) = \frac{1}{2}Vj\left(\frac{2}{j} + \frac{2}{k} - 1\right).$$

Proof We count the number of edges in the regular subdivision in two ways.

We first count the edges surrounding the faces. Each face has k edges, which suggests that there are Fk edges altogether — but each edge is counted twice (see the example of a tetrahedron in Figure 1.6), because either the edge belongs to two different faces and is counted once for each face, or the face meets itself along the edge and the edge is counted twice when we count the edges around that face. It follows that there are $\frac{1}{2}Fk$ edges in total, and so

$E = \frac{1}{2}Fk$.

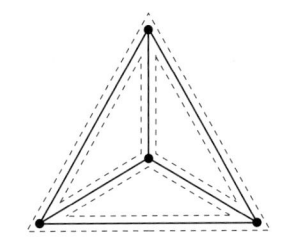

Figure 1.6

We next look at the edges incident to the vertices. Each vertex has j incident edges, which suggests that there are Vj edges altogether — but each edge is counted twice (see the example of a tetrahedron in Figure 1.7), because either the edge joins two different vertices and is counted once for each vertex, or the edge starts and finishes at the same vertex and is counted twice when we count the edges at that vertex. It follows that there are $\frac{1}{2}Vj$ edges in total, and so

$E = \frac{1}{2}Vj$.

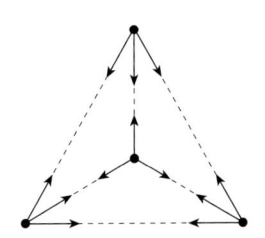

Figure 1.7

Rearranging these two equations, we obtain $F = 2E/k$ and $V = 2E/j$. Therefore

$$\chi = V - E + F = \frac{2E}{j} - E + \frac{2E}{k} = E\left(\frac{2}{j} + \frac{2}{k} - 1\right).$$

This is the first of the regular subdivision formulas.

The other two formulas are obtained from the first by replacing E in turn by $\frac{1}{2}Fk$ and $\frac{1}{2}Vj$. ∎

Problem 1.1

Verify the regular subdivision formulas when S is a sphere and the regular subdivision corresponds to each of the five regular polyhedra.

Recall that $\chi = 2$ for a sphere.

Our main aim in this section is to determine when a given surface without boundary and with given Euler characteristic has regular subdivisions — and when it has, to describe them by specifying which values of j, k, V, E and F are possible. The regular subdivision formulas have a key role to play in this, as we now begin to demonstrate.

When $\chi = 0$, it follows from the regular subdivision formulas that a regular subdivision can exist only when there are integers $j\,(\geq 2)$ and $k\,(\geq 2)$ such that

$$0 = \frac{2}{j} + \frac{2}{k} - 1 \quad \text{or, equivalently,} \quad 2j + 2k - jk = 0.$$

When $\chi \neq 0$, we can rewrite the formulas (after a little algebra) in the following form.

Corollary 1.2

Let S be a surface without boundary with Euler characteristic $\chi\,(\neq 0)$ and having a regular subdivision with V vertices, E edges and F faces, in which each face has k edges and each vertex has j incident edges. Then

$$V = \frac{2\chi k}{2j + 2k - jk}, \quad E = \frac{\chi jk}{2j + 2k - jk}, \quad F = \frac{2\chi j}{2j + 2k - jk}.$$

It follows that a regular subdivision can exist only when the expressions in Corollary 1.2 are all positive integers.

Problem 1.2

Prove Corollary 1.2.

The regular subdivision formulas provide only *necessary* conditions for a regular subdivision to exist on the given surface: just because the formulas are satisfied for particular values of j, k, V, E and F is no guarantee that there exists a regular subdivision with those values. However, these conditions also turn out to be *sufficient*, except in the case of the projective plane: that is to say, we have the following result.

> ### Theorem 1.3
> For a surface without boundary that is not a projective plane, suppose that there are integer values of $j\ (\geq 2)$ and $k\ (\geq 2)$ for which the values of V, E and F obtained from the regular subdivision formulas are positive integers. Then there exists a regular subdivision with these values of j, k, V, E and F.

This sufficiency result was proved as recently as 1980, much later than most of the results in this course. We omit its proof, as it uses techniques beyond the scope of this course.

Remark

When $\chi \neq 0$, we use the regular subdivision formulas in Corollary 1.2.
When $\chi = 0$, we use the regular subdivision formulas
$E = \frac{1}{2}Fk = \frac{1}{2}Vj$.

So our task, so far as this section is concerned, becomes that of analysing the regular subdivision formulas.

1.2 Dual subdivisions

Suppose we take a cube, place a new vertex in the middle of each face, and then join by an edge those new vertices that lie in neighbouring faces. We obtain an octahedron, as illustrated in Figure 1.8.

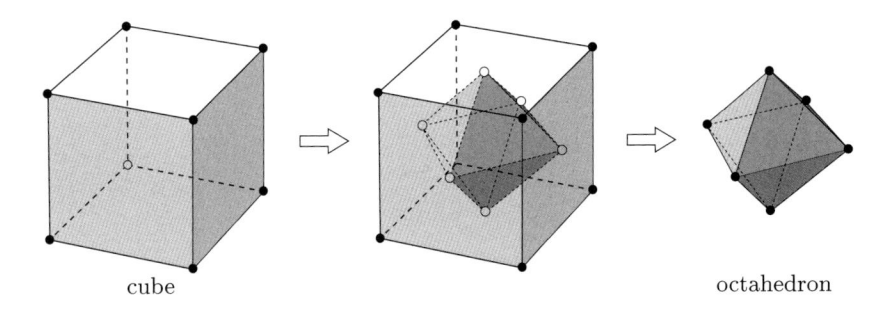

Figure 1.8

We can then repeat this construction with the octahedron to obtain a cube, as shown in Figure 1.9.

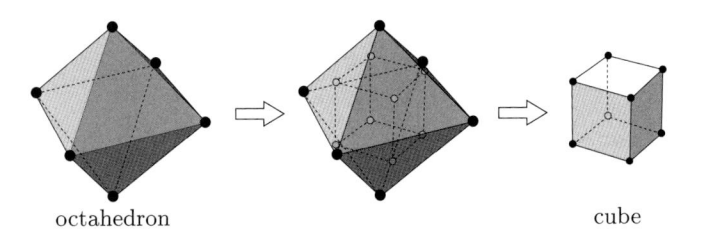

Figure 1.9

We say that the octahedron and cube are *dual polyhedra*. In a similar way, we can show that the dodecahedron and icosahedron are dual polyhedra.

Problem 1.3

What happens if you carry out this construction for a tetrahedron?

Thinking of these polyhedra as subdivisions of the sphere motivates the following general definition.

Definition

Given any subdivision S of a surface, there is another subdivision S^*, called a **dual subdivision**, which is constructed as follows:
- place a new vertex inside each face of the subdivision S — these are the vertices of S^*;
- for each edge e of S, draw a line joining the vertices of S^* in the faces on either side of e — these lines are the edges of S^*.

Each new edge should be drawn so that it crosses the original edge.

We illustrate this procedure, in Figure 1.10, for a subdivision of a surface without boundary.

Notice that, as it is a surface without boundary, there is an 'exterior' face in which we must place a new vertex.

subdivision S dual subdivision S^*

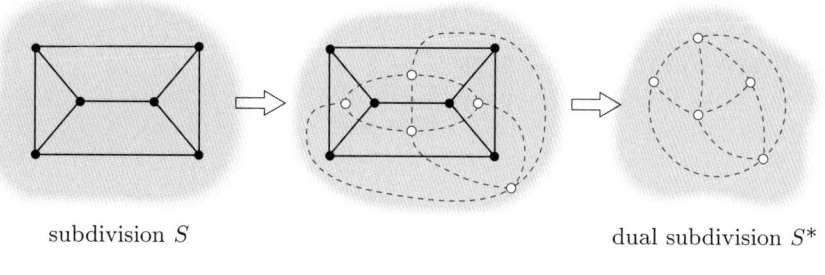

Figure 1.10

Problem 1.4

Draw the dual subdivision S^* of each of the subdivisions S of the surface without boundary in Figure 1.11.

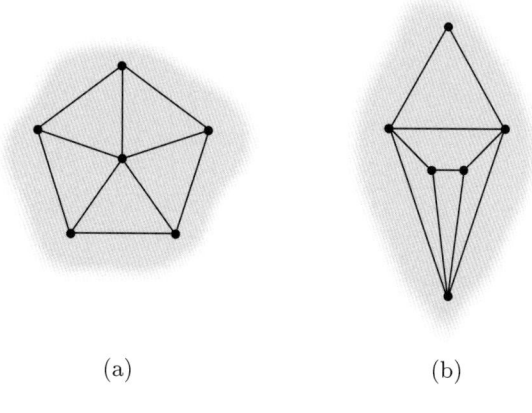

(a)   (b)

Figure 1.11

Notice that, for a subdivision S and its dual S^*:
- each vertex of S lies in exactly one face of S^*, and each face of S^* encloses exactly one vertex of S;
- each edge of S^* crosses exactly one edge of S, and each edge of S is crossed by exactly one edge of S^*, so the number E of edges is unchanged.

It follows that:
- if S has V vertices, E edges and F faces, then S^* has F vertices, E edges and V faces.

We deduce the following theorem.

> **Theorem 1.4 Existence of dual subdivisions**
> (a) Given a subdivision S with V vertices, E edges and F faces, there is a dual subdivision S^* with F vertices, E edges and V faces.
> (b) When S is a regular subdivision with k-sided polygons and j incident edges at each vertex, the dual S^* is a regular subdivision with j-sided polygons and k incident edges at each vertex.

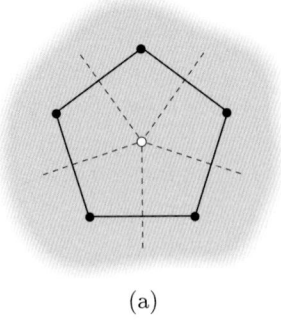

Proof
(a) This follows from the definition of a subdivision and the discussion above.
(b) When each face of S is a k-sided polygon, the k edges of S^* that enter a face of S meet at a vertex of S^* — so S^* has k incident edges at each vertex (see Figure 1.12(a)).

When S has j incident edges at each vertex, the face of S^* enclosing that vertex of S must have j edges — so each face of S^* is a j-sided polygon (see Figure 1.12(b)). ∎

We shall find dual subdivisions useful in what follows.

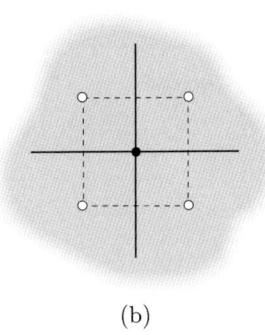

Figure 1.12

1.3 Regular subdivisions of the sphere

We now find all the regular subdivisions of the sphere, in which each face is a k-sided polygon and where each vertex has j incident edges.

When $k \geq 3$ and $j \geq 3$, we obtain the subdivisions corresponding to the regular polyhedra.

> **Theorem 1.5 The regular polyhedra**
> On the sphere, the regular subdivisions with $k \geq 3$ and $j \geq 3$ correspond to the regular polyhedra:
> - $k = 3$, $j = 3$ gives the *tetrahedron*: $V = 4$, $E = 6$, $F = 4$;
> - $k = 3$, $j = 4$ gives the *octahedron*: $V = 6$, $E = 12$, $F = 8$;
> - $k = 3$, $j = 5$ gives the *icosahedron*: $V = 12$, $E = 30$, $F = 20$;
> - $k = 4$, $j = 3$ gives the *cube*: $V = 8$, $E = 12$, $F = 6$;
> - $k = 5$, $j = 3$ gives the *dodecahedron*: $V = 20$, $E = 30$, $F = 12$.

Proof The Euler characteristic of the sphere is $\chi = 2$, and so Corollary 1.2 gives the following expressions for V, E and F:
$$V = \frac{4k}{2j + 2k - jk}, \quad E = \frac{2jk}{2j + 2k - jk}, \quad F = \frac{4j}{2j + 2k - jk}.$$
Since $k \geq 3$, we now look in turn at the cases $k = 3$, $k = 4$, $k = 5$ and $k \geq 6$.

When $k = 3$ (every face is a triangle), these formulas give
$$V = \frac{12}{6 - j}, \quad E = \frac{6j}{6 - j}, \quad F = \frac{4j}{6 - j}.$$
Since these must all be positive, $6 - j > 0$, so $j = 3$, 4 or 5. *Recall that we are interested only in the case where $j \geq 3$.*

- When $j = 3$, $V = \frac{12}{3} = 4$, $E = \frac{18}{3} = 6$, $F = \frac{12}{3} = 4$, giving the *tetrahedron*.
- When $j = 4$, $V = \frac{12}{2} = 6$, $E = \frac{24}{2} = 12$, $F = \frac{16}{2} = 8$, giving the *octahedron*.
- When $j = 5$, $V = \frac{12}{1} = 12$, $E = \frac{30}{1} = 30$, $F = \frac{20}{1} = 20$, giving the *icosahedron*.

When $k = 4$ (every face is a quadrilateral), these formulas give
$$V = \frac{8}{4 - j}, \quad E = \frac{4j}{4 - j}, \quad F = \frac{2j}{4 - j}.$$
Since these must all be positive, $4 - j > 0$, so $j = 3$.

- When $j = 3$, $V = \frac{8}{1} = 8$, $E = \frac{12}{1} = 12$, $F = \frac{6}{1} = 6$, giving the *cube*.

When $k = 5$ (every face is a pentagon), these formulas give
$$V = \frac{20}{10 - 3j}, \quad E = \frac{10j}{10 - 3j}, \quad F = \frac{4j}{10 - 3j}.$$
Since these must all be positive, $10 - 3j > 0$, so $j = 3$.

- When $j = 3$, $V = \frac{20}{1} = 20$, $E = \frac{30}{1} = 30$, $F = \frac{12}{1} = 12$, giving the *dodecahedron*.

Finally, when $k \geq 6$ and $j \geq 3$, we use some algebra to write
$$2j + 2k - jk = -4(j - 3) - (k - 6) - (j - 3)(k - 6) \leq 0.$$
Thus the denominator is never positive, so these cases cannot arise. It follows that the only values of V, E and F that can arise are those corresponding to the regular polyhedra. ∎

We can also investigate the cases where $k = 2$ (each face is a digon), and those where $j = 2$ (exactly two edges meet at each vertex). *A *digon* is a two-sided polygon.*

When $k = 2$, and for any j, say $j = n$, Corollary 1.2 gives
$$V = 2, \; E = n, \; F = n.$$
Thus, we have n digons that meet at two vertices, giving a picture resembling the segments of an orange (see Figure 1.13(a)).

When $j = 2$, and for any k, say $k = n$, Corollary 1.2 gives
$$V = n, \; E = n, \; F = 2.$$
Thus, there are two faces, each with n vertices and n edges, and we can imagine the faces as the northern and southern hemispheres of a globe, with the n vertices and the n edges running around the equator (see Figure 1.13(b)).

(a) orange   (b) globe

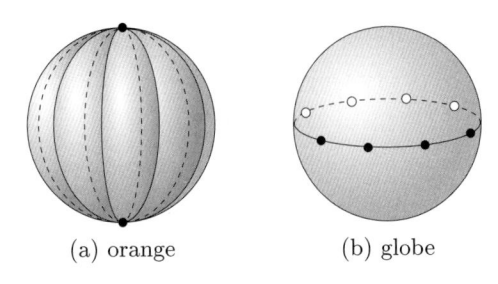

Figure 1.13

Problem 1.5

Show that the subdivisions in Figure 1.13 are dual subdivisions of the sphere.

There is an alternative approach to finding subdivisions, which will be useful when we consider subdivisions of surfaces other than the sphere. For the sphere, let us return to the formulas for V, E and F:

$$V = \frac{4k}{2j + 2k - jk}, \quad E = \frac{2jk}{2j + 2k - jk}, \quad F = \frac{4j}{2j + 2k - jk}.$$

We can determine the integers $j\,(\geq 2)$ and $k\,(\geq 2)$ for which $2j + 2k - jk > 0$ by drawing the curve $2j + 2k - jk = 0$, or equivalently the curve $k = 2j/(j - 2)$, in the positive quadrant of the jk-plane. This curve is shown in Figure 1.14, and divides the quadrant into two regions $2j + 2k - jk > 0$ in the shaded lower-left region, and $2j + 2k - jk < 0$ in the upper-right region. The curve is part of a rectangular hyperbola, symmetrical about the line $j = k$, whose asymptotes are the lines $j = 2$ and $k = 2$.

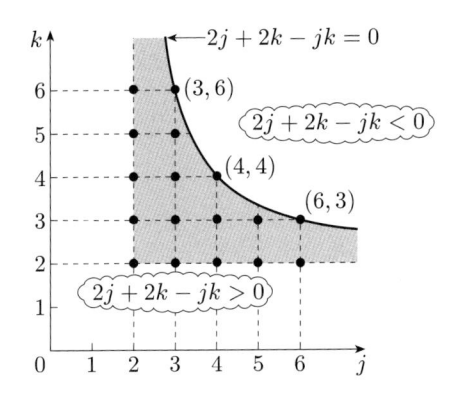

Figure 1.14

We wish to know which points with integer coordinates (j, k), with $j \geq 2$ and $k \geq 2$, lie in the shaded region, where $2j + 2k - jk > 0$. Now the curve $2j + 2k - jk = 0$ crosses the line $j = 3$ where $k = 6$, crosses the line $j = 4$ where $k = 4$, and crosses the line $j = 6$ where $k = 3$; it does not cross the line $j = 5$ at an integer value of k; it does not cross the asymptotes $j = 2$ and $k = 2$ at all. Taking these facts into account, we see that the points with integer coordinates in the shaded region (strictly below the curve) are:

- $(3, 3)$, $(3, 4)$, $(3, 5)$, $(4, 3)$ and $(5, 3)$;
- all points of the form $(n, 2)$, with n an integer and $n \geq 2$;
- all points of the form $(2, n)$, with n an integer and $n \geq 2$.

The first set of points corresponds to the five regular polyhedra, and the other two sets correspond to the orange and the globe. These are therefore the *only* regular subdivisions of the sphere, in the sense that *every* regular subdivision of the sphere is homeomorphic to one of them. It is a striking fact that we can draw these subdivisions in such a way that the faces are all congruent — and this is the case for other surfaces also.

1.4 Regular subdivisions of the torus

We next use the regular subdivision formulas to study regular subdivisions of the torus. In what follows, recall that
- each face has the same number k of edges;
- each vertex has the same number j of incident edges.

The Euler characteristic of the torus is $\chi = 0$, so the regular subdivision formulas imply that

$$0 = \frac{2}{j} + \frac{2}{k} - 1 \quad \text{or, equivalently,} \quad 2j + 2k - jk = 0.$$

We saw at the end of the previous subsection that the points with positive integer coordinates on this curve are $(3,6)$, $(4,4)$ and $(6,3)$. These give regular subdivisions of the torus, as we now show.

Theorem 1.6

The possible regular subdivisions of the torus are given in the following table, where n denotes an arbitrary positive integer.

j	k	V	E	F
3	6	$2n$	$3n$	n
4	4	n	$2n$	n
6	3	n	$3n$	$2n$

Proof As noted above, since the Euler characteristic of the torus is $\chi = 0$, regular subdivisions of the torus can only involve vertices with j incident edges and k-sided faces where $(j,k) = (3,6)$, $(4,4)$ or $(6,3)$.

Since the regular subdivision formulas involving χ all reduce to $0 = 0$ with the given choices of j and k, we need to use the formulas $E = \frac{1}{2}Fk = \frac{1}{2}Vj$.

- When $j = 3$, $k = 6$, we have $E = 3F$ and $2E = 3V$, and any positive integer values of V, E and F satisfying these relationships give a regular subdivision. Choosing F arbitrarily, say $F = n$, we obtain $E = 3n$ and $V = 2n$.
- When $j = k = 4$, we have $E = 2F = 2V$: choosing $F = n$, we obtain $E = 2n$ and $V = n$.
- When $j = 6$, $k = 3$, we have $2E = 3F$ and $E = 3V$: choosing $V = n$, we obtain $E = 3n$ and $F = 2n$.

By Theorem 1.3, each of these possibilities can occur, for all positive integers n. ∎

Alternatively, we can deal with the case $j = 6$, $k = 3$ by taking the dual of the subdivision with $j = 3$, $k = 6$.

Figure 1.15 illustrates each of the three types of subdivision of the torus, both in three-dimensional form and as rectangles with edge identifications.

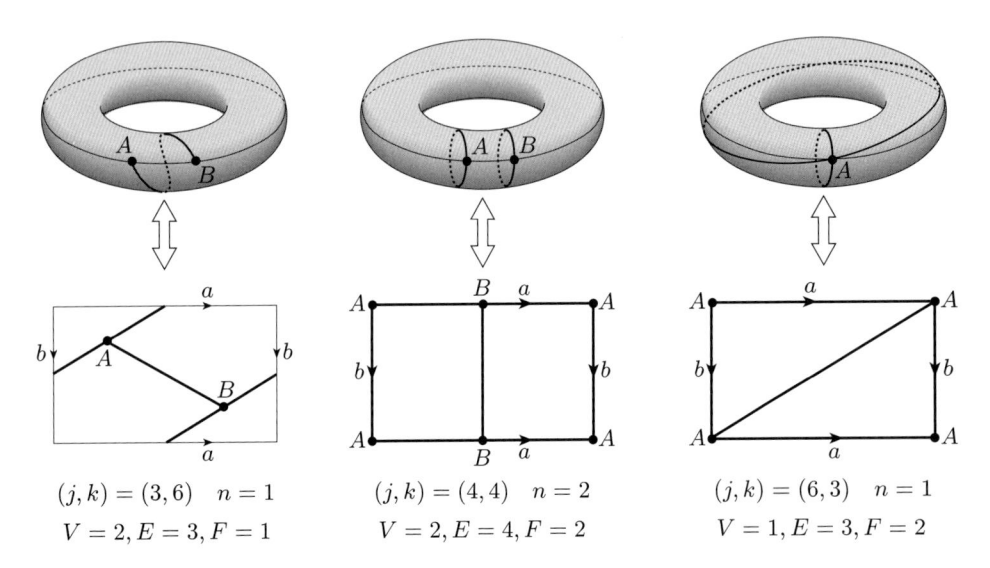

Note that the first and third of these regular subdivisions are dual to each other, and the second one is its own dual.

$(j, k) = (3, 6)$ $n = 1$
$V = 2, E = 3, F = 1$

$(j, k) = (4, 4)$ $n = 2$
$V = 2, E = 4, F = 2$

$(j, k) = (6, 3)$ $n = 1$
$V = 1, E = 3, F = 2$

Figure 1.15

Note that only the thicker lines in Figure 1.15 represent edges. Thus, in the case where $(j, k) = (3, 6)$, the boundary lines in the rectangular representation are not edges of the subdivision (though they are in the other two cases). From now on, the boundary lines in a polygonal representation of a subdivision of a surface may or may not be edges of the subdivision. Thicker lines will be used to indicate edges.

Problem 1.6

Draw diagrams of regular subdivisions on rectangular representations of the torus corresponding to the case $n = 3$.

1.5 Surfaces with negative Euler characteristic

We now turn our attention to finding subdivisions of surfaces without boundary for which $\chi < 0$.

First, we show that such a surface must have at least one regular subdivision. For convenience, we rewrite the formulas of Corollary 1.2 in the form

In fact, we shall see that such a surface must have at least three regular subdivisions.

$$V = \frac{2(-\chi)k}{jk - 2j - 2k}, \quad E = \frac{(-\chi)jk}{jk - 2j - 2k}, \quad F = \frac{2(-\chi)j}{jk - 2j - 2k}.$$

Since $-\chi > 0$, these formulas tell us that we need $jk - 2j - 2k > 0$, or equivalently $2j + 2k - jk < 0$. So, with reference to Figure 1.16, we are now working in the region above and to the right of the curve $2j + 2k - jk = 0$.

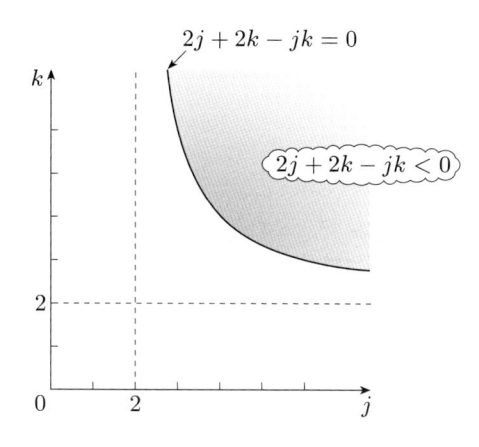

Figure 1.16

Now, this region contains the point $(3, 7)$, for which

$$jk - 2j - 2k = 21 - 6 - 14 = 1.$$

We deduce the following result.

Theorem 1.7

Every surface without boundary and with negative Euler characteristic has a regular subdivision with $j = 3$ and $k = 7$.

Proof When $j = 3$ and $k = 7$, $jk - 2j - 2k = 1$, and so, by Corollary 1.2,

$$V = 14(-\chi), \quad E = 21(-\chi), \quad F = 6(-\chi).$$

Therefore V, E and F are all positive integers. Hence, by Theorem 1.3, whatever the value of $\chi < 0$, the surface has a regular subdivision. ∎

The case of the projective plane, with $\chi = 1$, does not arise here.

Problem 1.7

Show that every surface without boundary and with negative Euler characteristic has:

(a) a regular subdivision with $j = 7$ and $k = 3$;
(b) a regular subdivision with $j = k = 5$.

Theorem 1.7 and Problem 1.7 thus tell us that a surface without boundary and with negative Euler characteristic must have at least three regular subdivisions.

For the rest of this section, we investigate two questions for a given surface without boundary and with negative Euler characteristic.
- How can we find all the regular subdivisions?
- Which regular subdivisions have faces with a given number of edges?

Using Theorem 1.3, we need only look for positive integer solutions to the regular subdivision formulas in Corollary 1.2, or more usually their reformulations at the start of this subsection. The following worked problem illustrates how this can be done.

Worked problem 1.1

Determine whether there exists a regular subdivision of a surface without boundary, with Euler characteristic $\chi = -4$, whose faces are hexagons and which has four edges incident at each vertex.

We asked this question in the Introduction.

Solution

With the values $k = 6$ and $j = 4$, we have

$$jk - 2j - 2k = 24 - 8 - 12 = 4,$$

and the regular subdivision formulas give

$$V = \tfrac{48}{4} = 12, \; E = \tfrac{96}{4} = 24, \; F = \tfrac{32}{4} = 8.$$

Since V, E and F are all positive integers, there is such a regular subdivision, by Theorem 1.3. ∎

Problem 1.8

Determine whether there exists a regular subdivision of a surface without boundary, with Euler characteristic $\chi = -3$, whose faces are pentagons and which has four edges incident at each vertex.

When dealing with regular subdivisions of the sphere and torus, we found it relatively straightforward to deal with the positive integer points on or below, and to the left of, the curve $2j + 2k - jk = 0$. This is not the case for the infinite region above and to the right of the curve. We can nevertheless develop a systematic way of finding all possible regular subdivisions in such a case, as we shall illustrate.

We first show that in each case we have only a finite number of regular subdivisions to consider.

Theorem 1.8

A surface without boundary and with negative Euler characteristic has a finite number of regular subdivisions.

Proof From Theorem 1.7 we know that the regular subdivision formulas have at least one solution ($j = 3, k = 7$).

We also know from Corollary 1.2 that we must have $jk - 2j - 2k > 0$ for any solution (j, k). But for $j = 2$, we have

$$jk - 2j - 2k = -2j < 0, \text{ whatever the value of } k \geq 2.$$

So we can assume that $j \geq 3$, and by duality (Theorem 1.4) we can also assume that $k \geq 3$.

Suppose first that $k \geq j$.

Now, for any subdivision, we must have $F \geq 1$, and so by Corollary 1.2,

$$\frac{2(-\chi)j}{jk - 2j - 2k} \geq 1.$$

Since $jk - 2j - 2k$ must be positive, we deduce that

$$2(-\chi)j \geq jk - 2j - 2k,$$

which we can rearrange as

$$k(j - 2) \leq 2(1 - \chi)j.$$

Since $j - 2 > 0$, we have

$$k \leq \frac{2(1 - \chi)j}{j - 2}. \tag{1.1}$$

So, under our assumption that $k \geq j$, we have
$$j \leq \frac{2(1-\chi)j}{j-2}.$$
It follows that
$$\frac{2(1-\chi)}{j-2} \geq 1,$$
so that $j - 2 \leq 2(1-\chi)$, and hence
$$j \leq 4 - 2\chi. \tag{1.2}$$

For a given value of $\chi < 0$, there are only finitely many positive integers j that satisfy (1.2), and for each such integer j there are only finitely many positive integers k that satisfy (1.1). So there are finitely many solutions with $k \geq j$.

By repeating the above arguments for the case where $j \geq k$, or by duality (Theorem 1.4), we can deduce that there are also finitely many solutions with $j \geq k$.

Thus, there is a finite number of solutions altogether. ∎

Figure 1.17 shows shaded the finite region of the jk-plane in which the integer points corresponding to a regular subdivision must lie.

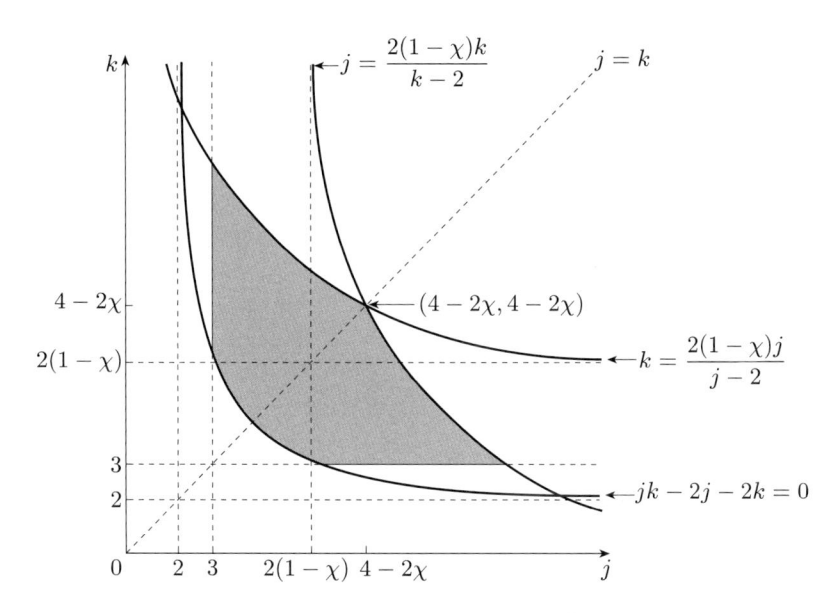

Figure 1.17 Region of possible integer points for $\chi < 0$

We do not assert that every positive integer point in the region indicated in Figure 1.17 gives a regular subdivision, but only that those points that do give a regular subdivision lie in the region.

The above proof suggests the following systematic method of searching for solutions. This method produces them all, given sufficient time.

> *Finding regular subdivisions of surfaces with $\chi < 0$*
>
> Every possible regular subdivision of a surface without boundary and with Euler characteristic $\chi < 0$ can be obtained by the following process.
>
> 1. Take each integer j such that
> $$3 \leq j \leq 4 - 2\chi$$
> in turn and, for each such value of j, list each integer k such that
> $$j \leq k \leq \frac{2(1-\chi)j}{j-2}.$$
>
> 2. For each pair (j, k) found in Step 1, check whether the values of V, E and F, given by Corollary 1.2, are positive integers: each pair for which they are all positive integers corresponds to a regular subdivision.
>
> 3. For each pair (j, k) found in Step 2 to correspond to a regular subdivision, there is also a regular subdivision corresponding to the pair (k, j).

This follows by Theorem 1.3.

Step 3 follows by duality (Theorem 1.4).

Remark

Each pair (j, k) tells us that the subdivision has k-sided faces and vertices incident to j edges. Substituting the j and k values into the regular subdivision formulas in Corollary 1.2 gives values for V, E and F, the numbers of vertices, edges and faces in the subdivision.

We now illustrate this process by considering regular subdivisions on the 2-fold torus, for which $\chi = -2$.

Worked problem 1.2

Find all the regular subdivisions of the 2-fold torus.

Solution

Since $\chi = -2$, we have $4 - 2\chi = 8$, and so $3 \leq j \leq 8$: we must therefore consider the cases $j = 3, 4, 5, 6, 7, 8$.

When $j = 3$, we have $\frac{2(1-\chi)j}{j-2} = 18$;

thus, the range of values of k is $3 \leq k \leq 18$.

The regular subdivision formulas give

$$V = \frac{4k}{k-6}, \quad E = \frac{6k}{k-6}, \quad F = \frac{12}{k-6}.$$

We could now take each value of k in the given range and determine whether these expressions all take positive integer values. However, we can be smarter than this. We note that, for F to be a positive integer, $k - 6$ must be a factor of 12. Since the factors of 12 are 1, 2, 3, 4, 6 and 12, k can take only the values 7, 8, 9, 10, 12 and 18. V and E are both integers for all these values of k, and we obtain the table of regular subdivisions for $j = 3$, with $k \geq j$, shown on the left below.

We consider $j = 3, 7, 8$ here and ask you to consider $j = 4, 5, 6$ in Problem 1.9.

Corollary 1.2.

By duality, we also deduce the existence of the dual subdivisions, with $k = 3$ and $j \geq k$, listed in the table on the right below.

j	k	V	E	F
3	7	28	42	12
3	8	16	24	6
3	9	12	18	4
3	10	10	15	3
3	12	8	12	2
3	18	6	9	1

j	k	V	E	F
7	3	12	42	28
8	3	6	24	16
9	3	4	18	12
10	3	3	15	10
12	3	2	12	8
18	3	1	9	6

When $j = 7$, we have $\dfrac{2(1-\chi)j}{j-2} = \dfrac{42}{5}$;

thus, the range of values of k is $7 \leq k \leq 8$, so $k = 7$ or $k = 8$.

The regular subdivision formulas give

$$V = \frac{4k}{5k-14}, \quad E = \frac{14k}{5k-14}, \quad F = \frac{28}{5k-14}.$$

But $5k - 14$ is not a factor of 28 when $k = 7$ or $k = 8$. So there are no regular subdivisions with $j = 7$ and $k \geq j$. Likewise, by duality, there are none with $k = 7$ and $j \geq k$.

When $j = 8$, we have $\dfrac{2(1-\chi)j}{j-2} = 8$;

so there is just the single value $k = 8$ to consider.

In this case the regular subdivision formulas give $V = 1$, $E = 4$, $F = 1$. This regular subdivision of the 2-fold torus can be represented as an octagon with edges identified in pairs (see Figure 1.18). ∎

Note that $\dfrac{2(1-\chi)j}{j-2}$ need not be an integer, so the upper limit on k is the largest integer that does not exceed $\dfrac{2(1-\chi)j}{j-2}$.

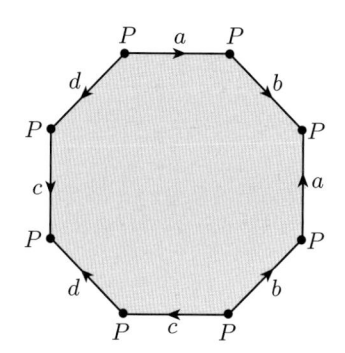

Figure 1.18

We saw the 2-fold torus represented in this way in Section 1 of *Unit B1*.

Problem 1.9

Complete the solution to Worked problem 1.2 by considering the cases where $j = 4, 5, 6$.

Problem 1.10

(a) Show that, for each surface without boundary and with negative Euler characteristic, there is a regular subdivision with $V = F = 1$, and find how many edges it has.

(b) How many edges are there in such a regular subdivision of an n-fold torus, where $\chi = 2 - 2n$?

Finding all the regular subdivisions of a given surface without boundary and with negative Euler characteristic can involve a lot of work. However, if the faces have a specified number of sides (that is, if the value of k is given), then the task may become much simpler, as we now illustrate.

By duality, the task may also become much simpler if the value of j is given.

Worked problem 1.3

Find all the regular subdivisions with pentagonal faces of a surface without boundary and with Euler characteristic $\chi = -3$.

Solution

In the previous examples we found all the possible values of k for a given value of j. Here we need to find the possible values of j for a given value of k ($k = 5$), and so we work with the dual formulation of the problem.

From the proof of Theorem 1.8 we can deduce that j must satisfy

$$3 \leq j \leq \frac{2(1-\chi)k}{k-2} = \frac{40}{3},$$

that is, $3 \leq j \leq 13$.

The regular subdivision formulas become

$$V = \frac{30}{3j-10}, \quad E = \frac{15j}{3j-10}, \quad F = \frac{6j}{3j-10}.$$

The smart move now is to consider V, and to find all the values of j in the given range for which $3j - 10$ is a factor of 30. The factors of 30 are 1, 2, 3, 5, 6, 10, 15 and 30, and the corresponding values of j are $\frac{11}{3}, \frac{12}{3}, \frac{13}{3}, \frac{15}{3}, \frac{16}{3}, \frac{20}{3}, \frac{25}{3}$ and $\frac{40}{3}$. Only two of these are integers: $j = \frac{12}{3} = 4$ and $j = \frac{15}{3} = 5$.

When $j = 4$, we have $V = 15$, $E = 30$, $F = 12$.

When $j = 5$, we have $V = 6$, $E = 15$, $F = 6$.

Since these values for V, E and F are all positive integers, both cases correspond to regular subdivisions, and these are the only regular subdivisions with pentagonal faces. ∎

Remark

The general technique for finding all the regular subdivisions with a given value for k for a surface without boundary and with negative Euler characteristic χ is to consider all j in the range

$$3 \leq j \leq \frac{2(1-\chi)k}{k-2}$$

and see which values give positive integer values for V, E and F in the regular subdivision formulas of Corollary 1.2.

Problem 1.11

Find all the regular subdivisions with pentagonal faces on a surface without boundary and with Euler characteristic $\chi = -6$.

$3 \leq j \leq \frac{2(1-\chi)k}{k-2} \qquad V = \frac{2(-\chi)k}{jk-2j-2k} \qquad E = \frac{(-\chi)jk}{jk-2j-2k} \qquad F =$

$3 \leq j \leq \frac{70}{3} \qquad\qquad = \frac{60}{3j-10} \qquad\qquad = \frac{30j}{3j-10} \qquad\qquad \frac{12j}{3j-10}$

$3 \leq j \leq 23$

factors of 60: 1, ②, 3, 4, ⑤, 6, 10, 12, 15, ㉔, 30,

$f = 3j - 10$

$j = \frac{f+10}{3}$

$\boxed{4, 5, 10}$

2 Finding the characteristic numbers

After working through this section, you should be able to:
▶ determine the characteristic numbers of a surface given as a polygon with edge identifications;
▶ decide whether a given surface is orientable by drawing thickened curves on the surface;
▶ explain how the orientability of a surface, described as a polygon with edge identifications, is determined by the directions of pairs of identified edges.

In *Unit B1* we stated that every compact surface can be represented as a polygon with edge identifications, and it is this method of representing a surface that we shall use when we prove the Classification Theorem in the next unit. For this purpose, we need to develop techniques for finding the three characteristic numbers of a polygon with edge identifications — the Euler characteristic χ, the boundary number β and the orientability number ω. This is the aim of this section.

We shall illustrate the techniques by reference to the compact surface given by the polygon with edge identifications in Figure 2.1. Recall that we assign to each edge a letter and an arrow. In Figure 2.1, we imagine that the edges marked a are identified in the senses indicated by the arrows, and the edges marked b and c are similarly identified. The edges marked d and e have no partners, but we still assign them arrows (in an arbitrary direction).

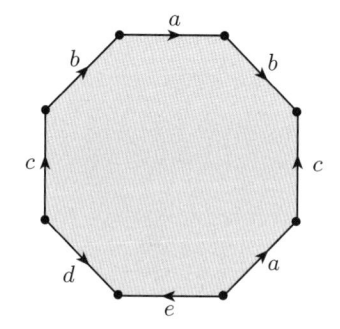

Figure 2.1

2.1 Finding the Euler characteristic

In order to find the Euler characteristic χ of a surface given as a polygon with edge identifications, we need to know the number of vertices V, the number of edges E and the number of faces F on the corresponding surface. We illustrate the ideas with reference to the example in Figure 2.1.

Faces There is only 1 face, the polygon itself, so $F = 1$.

Edges Each pair of identified edges (a, b and c) counts as a single edge on the surface. The edges that appear only once (d and e) correspond to edges on the boundary of the surface. So the total number of edges on the surface is the number of distinctly labelled edges of the polygon — here they are a, b, c, d and e, so $E = 5$.

Vertices To find the number of vertices requires some care, as we now illustrate. Our procedure is called the **method of inserting vertices**.

We begin by arbitrarily labelling all the vertices of the polygon, with a different symbol for each vertex, as in Figure 2.2.

Look first at the edges marked a. The edge a at the top starts at vertex P and finishes at Q. The other edge a starts at vertex T and ends at S. Since we are identifying the two edges marked a, their starting points must agree, and so must their endpoints. So, as vertices of the surface,

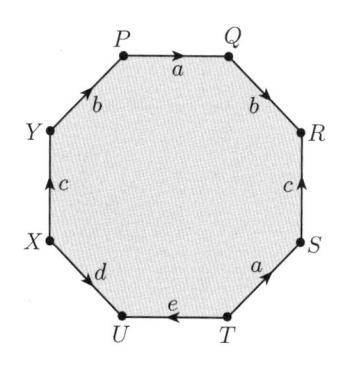

$$T = P \quad \text{and} \quad S = Q.$$

Figure 2.2

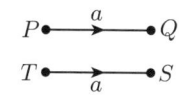

Figure 2.3

We look next at the two appearances of the edge b. One of them goes from Q to R, and the other goes from Y to P. Since we are identifying the edges marked b, their starting points and endpoints must agree. So, as vertices of the surface,

$Y = Q$ and $P = R.$

Figure 2.4

How about the two appearances of the edge c? Comparing the starting points and endpoints, we see that

$X = S$ and $Y = R.$

It follows that, when all the edge identifications have been made to obtain the surface, we have

$P = Q = R = S = T = X = Y.$

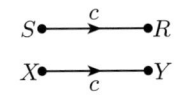

Figure 2.5

What about the vertex U? It appears between the edges d and e, and these edges are not identified with any other edges, so there can be no further identifications of vertices.

Thus, the surface has two distinct vertices (P and U), as illustrated in Figure 2.6, and so $V = 2$.

In practice, a quicker way of carrying out the above method of inserting vertices is to draw the polygon as in Figure 2.2 and then change the letters as you proceed. For example, after considering the edges labelled a, you can cross out T and S and replace them by P and Q, respectively. Then, after considering the edges labelled b, you can replace Y by Q and R by P. Finally, considering the edges labelled c, you can replace X by Q (formerly S) and Q (formerly Y) by P (formerly R). In this way, you obtain Figure 2.6.

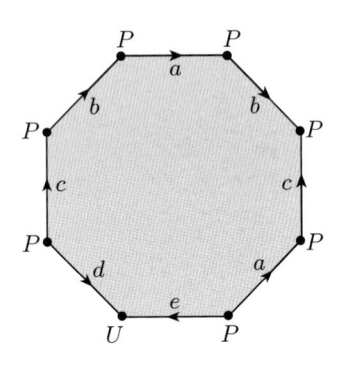

Figure 2.6

We conclude that the Euler characteristic χ of the surface is

$\chi = V - E + F = 2 - 5 + 1 = -2.$

Problem 2.1

Find the numbers V, E and F for the surface represented by the polygon with edge identifications in Figure 2.7, and hence find the Euler characteristic of the surface.

Problem 2.2

Find the Euler characteristic of each of the surfaces given by polygons with edge identifications in Figure 2.8.

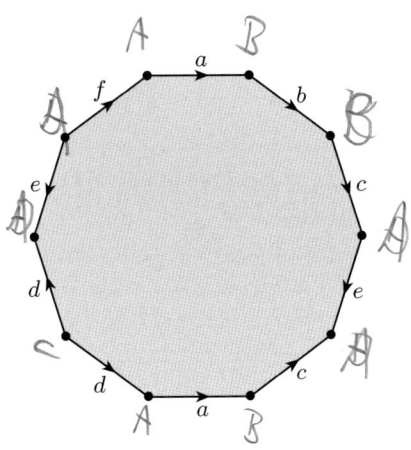

Figure 2.7

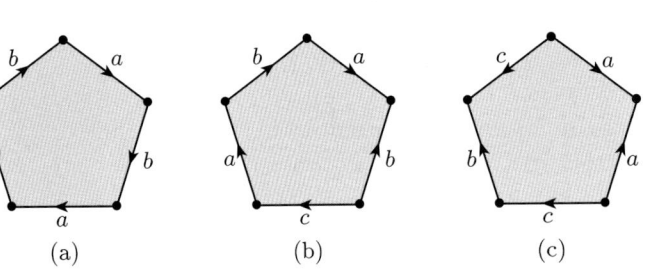

(a)   (b)   (c)

Figure 2.8

2.2 Finding the boundary number

What is the boundary number β of the surface in Figure 2.1?

The edges that occur only once around the polygon must form part of the boundary of the surface, so the question is: do the two boundary edges d and e form one connected piece or two pieces?

To answer this we look at the vertices incident with these edges. Here, referring to Figure 2.6, the edge d starts at vertex P and goes to U, and the edge e also starts at vertex P and goes to U. So the two boundary edges form a single connected piece, joined at the vertices P and U, as illustrated in Figure 2.9. Thus the boundary number β is 1.

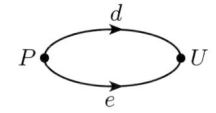

Figure 2.9

Problem 2.3

Find the boundary number of the surface in Problem 2.1.

Problem 2.4

Find the boundary number of each of the surfaces in Problem 2.2.

2.3 Finding the orientability number

What is the orientability number ω of the surface in Figure 2.1?

In *Unit B1* we saw that a surface is *orientable* if we can define 'clockwise' consistently at every point of the surface, that is, in such a way that all neighbouring clocks agree. The basic idea is that a clock-face lying in the surface at a point defines an orientation at that point.

> A surface is *orientable* ($\omega = 0$) if it can be oriented all over in a consistent manner, and *non-orientable* ($\omega = 1$) otherwise.

We also saw in *Unit B1* that a surface is non-orientable if and only if it contains a Möbius band. By a Möbius band in this context we mean a closed curve 'thickened' by tracing over it with a thick crayon, so that a region homeomorphic to a Möbius band is created on the surface. Recall that we identified such Möbius bands in surfaces by examining closed curves in the surface.

Unit B1, Theorem 2.2.

We shall use this idea, in a slightly different way, to determine the orientability of compact surfaces represented as polygons with edge identifications. We consider closed curves that are drawn on such polygons and define the *thickened neighbourhood* of such a curve to be the set of all points that lie no more than a fixed finite distance from it, where we pick a suitably small value of the distance to ensure that the thickened neighbourhood has no overlaps. An example is shown in Figure 2.10, which shows a thickened neighbourhood of a closed curve on a torus represented as a rectangle with edge identifications; the thickened neighbourhood includes the two dotted curves.

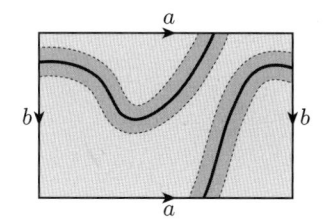

Figure 2.10

The thickened neighbourhood of the curve can be considered as a surface, and it can be shown that this surface can be homeomorphic only to a cylinder or to a Möbius band. We can tell which of these it is by looking at the boundary of the thickened neighbourhood, which must come in either one piece or two. When the boundary is in one piece, the thickened neighbourhood is homeomorphic to a Möbius band; when the boundary is in two pieces, the thickened neighbourhood is homeomorphic to a cylinder.

The proof, which uses the Jordan Curve Theorem (mentioned in *Unit B1*), lies beyond the scope of this course.

Recall that a Möbius band has one boundary component, while a cylinder has two.

If we can find a thickened neighbourhood of a closed curve that is homeomorphic to a Möbius band, then we know that the surface is non-orientable. If we can show that *all* closed curves are homeomorphic to cylinders, then the surface is orientable.

The idea of looking at the boundaries of thickened curves to check for Möbius bands will lead us to a simple technique for determining orientability. First, however, we show how to construct a thickened neighbourhood of a given closed curve and check the number of boundary components.

Figure 2.11(a) shows a polygon with edge identifications, representing a torus, which is orientable. We have marked on it a closed curve C from a point P to itself. We can draw the boundary of a thickened neighbourhood of the curve C by starting at a point Q near P on one edge and drawing a curve that stays close to C and runs to the same point Q on the opposite edge, as in Figure 2.11(b). We then choose another point R near P on one edge, but on the opposite side of P from Q, and draw another curve that runs close to C and runs to the same point R on the opposite edge, as in Figure 2.11(c). The result is a thickened neighbourhood of C with two boundary components (the curve through Q and the curve through R). The thickened neighbourhood of C is therefore a cylinder, as illustrated in Figure 2.11(d).

Remember that opposite edges of the rectangle are identified in *the same* direction.

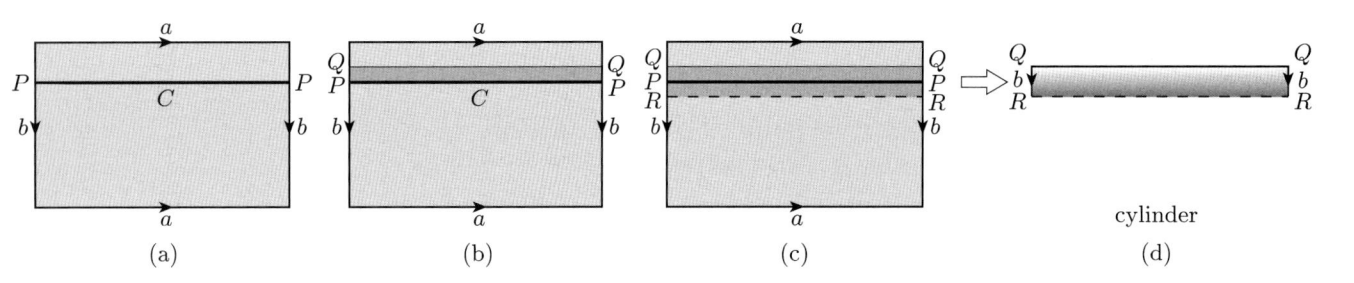

Figure 2.11

Figure 2.12(a) shows another polygon with edge identifications, representing a Klein bottle, which is non-orientable.
We have marked on it a closed curve C from P to itself, as before.

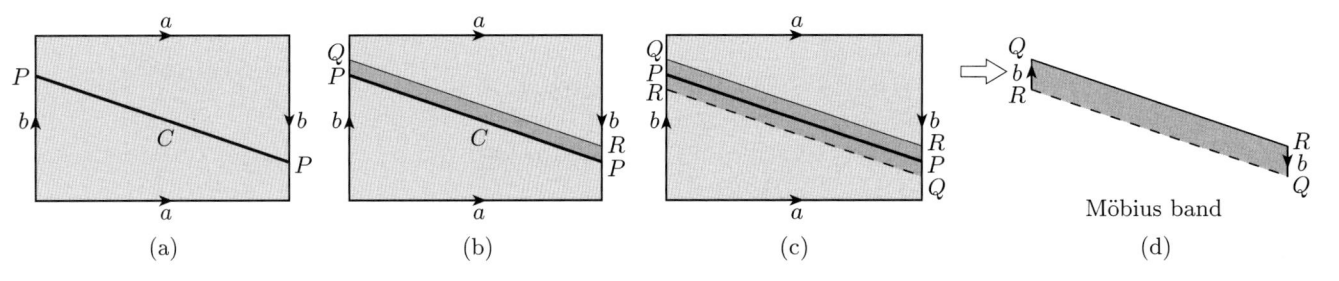

Figure 2.12

We draw a thickened neighbourhood of the curve C by starting at a point Q near to P on one edge and drawing a curve that stays close to C, and runs to a different point R on the opposite edge, as in Figure 2.12(b). We then locate the point on the left edge with which R is identified (also labelled R) and draw a curve that runs close to C and runs to the point Q on the opposite edge, as in Figure 2.12(c). The result is a thickened neighbourhood of C with just one boundary component. The thickened neighbourhood of C is therefore a Möbius band, as illustrated in Figure 2.12(d).

Remember that opposite edges of the rectangle are identified in *opposite* directions.

Worked problem 2.1

Draw a thickened neighbourhood of the curve C in Figure 2.13 and determine whether the neighbourhood is a cylinder or a Möbius band. (The curve C starts from an edge labelled a, proceeds via the edges labelled c, and ends back at the edges labelled a.) What can you deduce about the orientability of the surface?

Solution

The curve has a thickened neighbourhood that is a Möbius band, as Figure 2.14 shows. Thus, the surface is non-orientable.

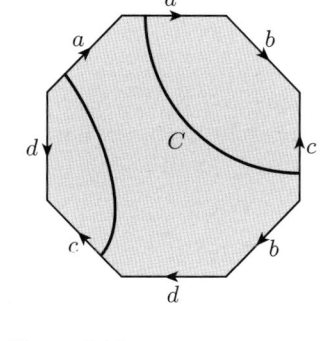

Figure 2.13

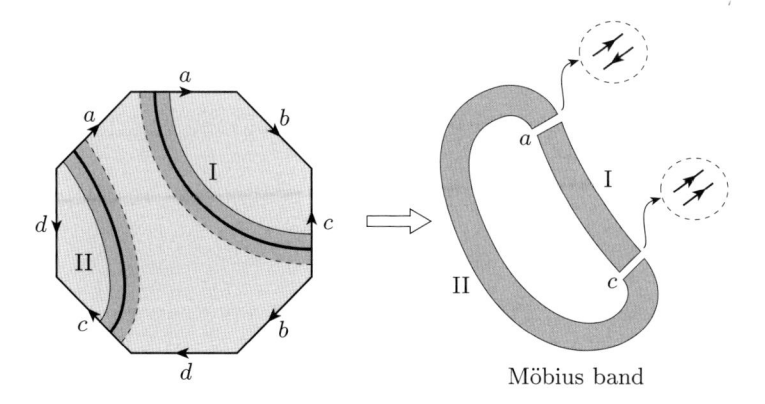

Figure 2.14 ∎

If we can find a Möbius band, as in Worked problem 2.1, then we know that the surface is non-orientable. However, remember that for an orientable surface we need *all* closed curves to lead to cylinders — finding a single cylinder does not imply that the surface is orientable. For example, consider the polygon with edge identifications shown in Figure 2.15(a). In Figure 2.15(b), we have joined corresponding points on the two edges labelled a by a curve C, and drawn a thickened neighbourhood of C. This has two boundary components, and so it is a cylinder. Now look at Figure 2.15(c), where corresponding points on the two edges labelled b are joined by a curve D. Here the thickened neighbourhood of D has just one boundary component, and is therefore a Möbius band. So, despite curve C giving a cylinder, the surface is non-orientable.

This is the example from Figure 2.1.

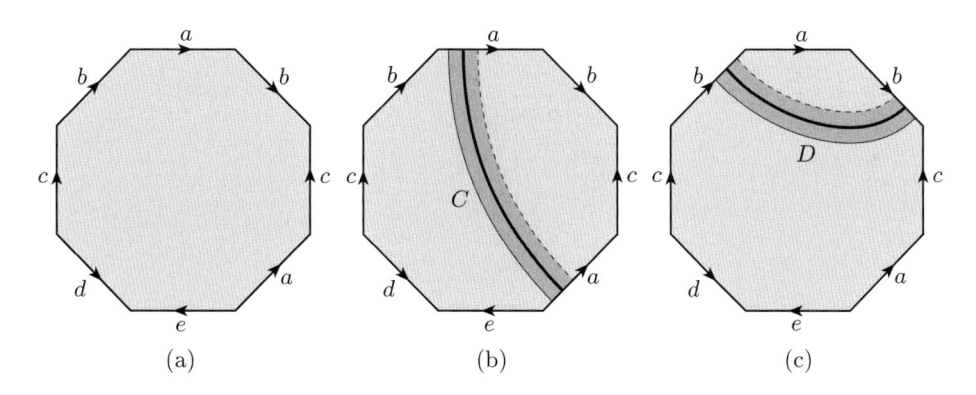

Figure 2.15

Problem 2.5

For each of the polygons in Figure 2.16, draw a thickened neighbourhood of the curve and determine whether the neighbourhood is a cylinder or a Möbius band. What conclusion, if any, can you draw about the orientability of each surface?

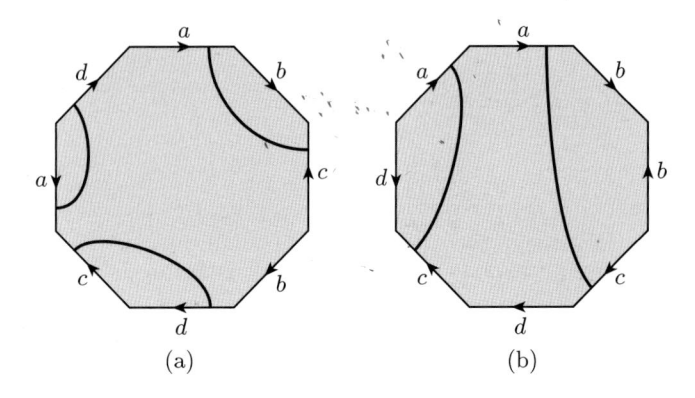

Figure 2.16

It may be clear to you by now that the relative directions of the arrows on the identified edges determine whether the thickened neighbourhood of a curve is a cylinder or a Möbius band. In order to express this notion of the 'direction of an arrow' precisely, we imagine going around the circumference of the polygon *clockwise*. We say that:

- two edges to be identified *occur in the same sense* if, when we go round the circumference of the polygon, their arrows both point in the direction of motion, or both point opposite to the direction of motion;
- two such edges *occur in opposite senses* if one arrow points in the direction of motion, and the other points opposite to it.

For example, in Figure 2.17, the edges labelled a occur in the same sense, and the edges labelled b occur in opposite senses.

Note that, although we may choose to go around the polygon in a clockwise direction, the definition does not depend on this choice. Furthermore, reversing the arrows on both members of a pair of identified edges makes no difference to whether they occur in the same sense or opposite senses.

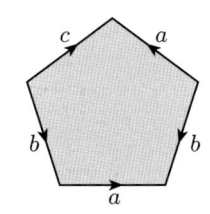

Figure 2.17

Figure 2.18 demonstrates that:

- by joining corresponding points on a pair of identified edges that occur *in opposite senses*, we obtain a closed curve whose thickened neighbourhood is a cylinder;
- by joining corresponding points on a pair of identified edges that occur *in the same sense*, we obtain a closed curve whose thickened neighbourhood is a Möbius band.

These may seem counter-intuitive at first.

Figure 2.18

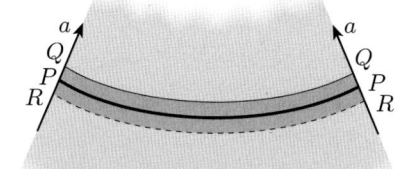

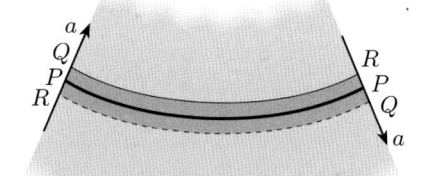

opposite senses | same sense

Worked problem 2.2

By considering the directions of the edges, determine whether the surface in Figure 2.19 is orientable.

Solution

The edges labelled a occur in the same sense, and so there is a closed curve whose thickened neighbourhood is a Möbius band. (The edges labelled b also occur in the same sense, giving another closed curve whose neighbourhood is a Möbius band.) Thus, the surface is non-orientable. ∎

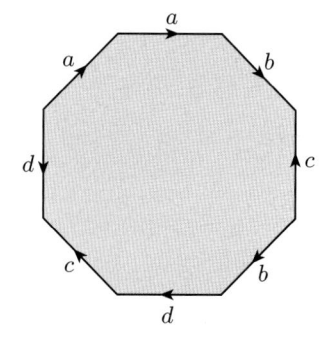

Figure 2.19

Problem 2.6

By considering the directions of the edges, determine whether each of the surfaces in Figure 2.20 is orientable.

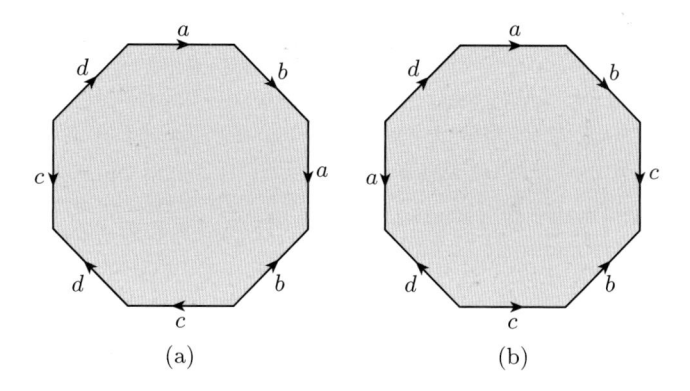

(a) (b)

Figure 2.20

We now prove a useful result that you may have already guessed.

Theorem 2.1

A surface defined by a polygon with edge identifications is non-orientable ($\omega = 1$) if and only if there is a pair of identified edges that occur in the same sense.

Proof Suppose that an edge occurs twice in the same sense in the polygon. Then the surface contains a Möbius band joining these edges, and is therefore non-orientable.

If you are short of time, you may wish to omit this proof on a first reading.

To prove the converse, we use a proof by contradiction. Consider a polygon in which each pair of identified edges occurs in both senses. We show that the surface cannot be non-orientable, by showing that there can be no closed curve that reverses orientation.

No closed curve C that reverses orientation can lie entirely inside the polygon, so we consider a curve C that crosses an edge of the polygon at a point P, say (see Figure 2.21).

Suppose that C leaves the polygon at a point on an edge a and rejoins it at the corresponding point on the other copy of the edge a. In Figure 2.21 we have drawn an oriented circle X (with an anticlockwise arrow), centred at a point of C, about to arrive at the edge a and leave the polygon.

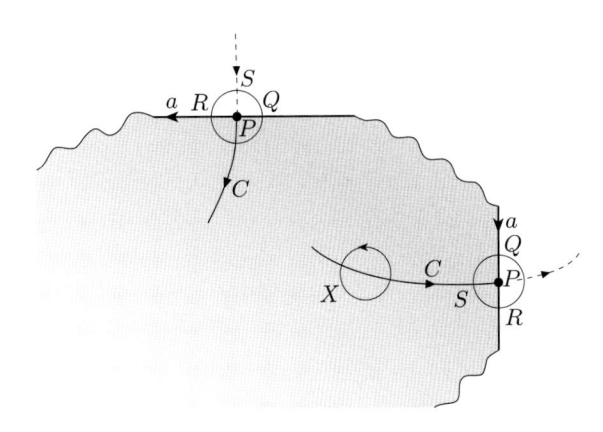

Figure 2.21

Let Q and R be the points where the circle meets the edge a, and let S be the point where the circle crosses C behind P (relative to the direction of motion along the curve). Note that the positions of Q, R and S on the circle specify its orientation: with the circle orientated anticlockwise, they appear in the order $QSRQ$.

Now locate the corresponding point P on the other edge a where the curve rejoins the polygon, and also the points corresponding to Q, R and S on the oriented circle centred at P. Because the edges have opposite senses, the points P, Q and R occur in the order shown on the second edge. The point S, being behind P, lies 'outside' the polygon, as shown. By observing the orientation of the upper circle determined by these points in the order $QSRQ$, we see that the circle reappears above with the same orientation as before. This shows that the circle on the second part of the curve is also orientated anticlockwise.

Thus, crossing an edge that occurs twice in opposite senses does not change the orientation of a circle centred on the curve. Therefore, if the edges of each identified pair occur in opposite senses, no closed curve can reverse orientation, and thus the surface cannot be non-orientable. ∎

Problem 2.7

Use Theorem 2.1 to determine whether the surface defined by each of the polygons with edge identifications in Figure 2.22 is orientable.

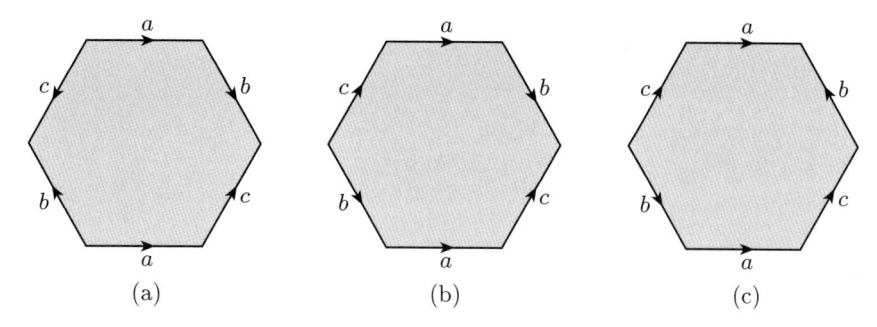

Figure 2.22

Problem 2.8

For each of the figures in Problem 2.2 (Figure 2.8), find the orientability number of the surface defined by the polygon with edge identifications.

3 Edge expressions

After working through this section, you should be able to:
- write down the *edge expression* and *edge equation* of a given polygon with edge identifications;
- draw the polygon corresponding to a given edge expression;
- determine the characteristic numbers of a surface from its description as an edge expression;
- use edge equations to convert operations on polygons with edge identifications into algebraic form.

In the previous section we discussed how to calculate the characteristic numbers of a compact surface represented as a polygon with edge identifications. We now show how it is possible to dispense with the polygon and work only with the list of edges.

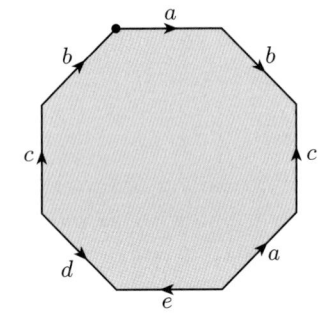

3.1 The idea of an edge expression

To introduce the ideas, we return to our example of Section 2 (Figure 2.1), the polygon with edge identifications shown in Figure 3.1. We list the edges in the order in which they occur as we go clockwise round the polygon, having arbitrarily picked an edge to start with. We use the following convention to distinguish sense (see Figure 3.2):
- when an edge x is traversed in the direction of its arrow, we write the edge label x;
- when an edge x is traversed in the opposite direction to that of the arrow on it, we write the edge label x^{-1}, with an exponent -1.

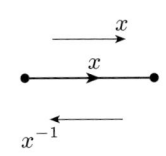

Thus, starting at the marked vertex at the top of the polygon in Figure 3.1, and going clockwise around the polygon, we obtain the list

$$abc^{-1}a^{-1}ed^{-1}cb.$$

With this convention, any surface represented as a polygon with edge identifications yields a corresponding list, called an **edge expression** for the surface.

Note that this convention agrees with algebraic usage. For example,
- the equation $(x^{-1})^{-1} = x$ corresponds to the fact that if we reverse the direction of an edge twice, we regain the original direction;
- the equation $(xy)^{-1} = y^{-1}x^{-1}$ corresponds to the fact that if we reverse the direction of 'x followed by y', we first traverse y backwards, and then traverse x backwards (see Figure 3.3).

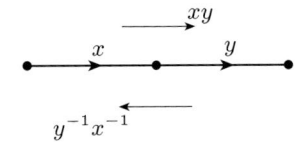

Figure 3.1

Figure 3.2

As we shall see, each surface can give rise to several possible edge expressions.

Figure 3.3

Problem 3.1

Starting from the vertex P in each case, write down the edge expression for each of the polygons with edge identifications in Figure 3.4.

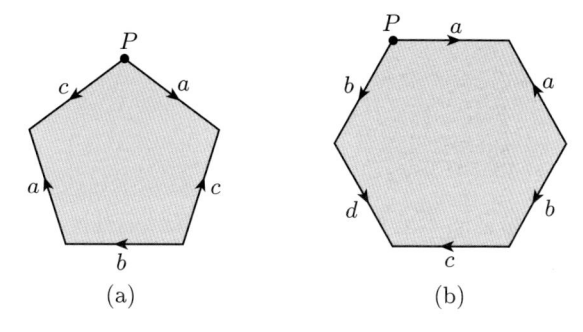

Figure 3.4

We can reverse the process. Given an edge expression, we can draw a corresponding polygon with edge identifications. The number of edges of the polygon is the number of symbols in the edge expression. So all we have to do is draw a polygon with the requisite number of edges, and then label and direct the edges in the way specified by the expression. For example, the edge expression

$$ab^{-1}ac^{-1}dc^{-1}ed$$

gives rise to the 8-sided polygon shown in Figure 3.5(a), while the edge expression

$$aabbcdc^{-1}efe^{-1}$$

gives rise to the 10-sided polygon shown in Figure 3.5(b).

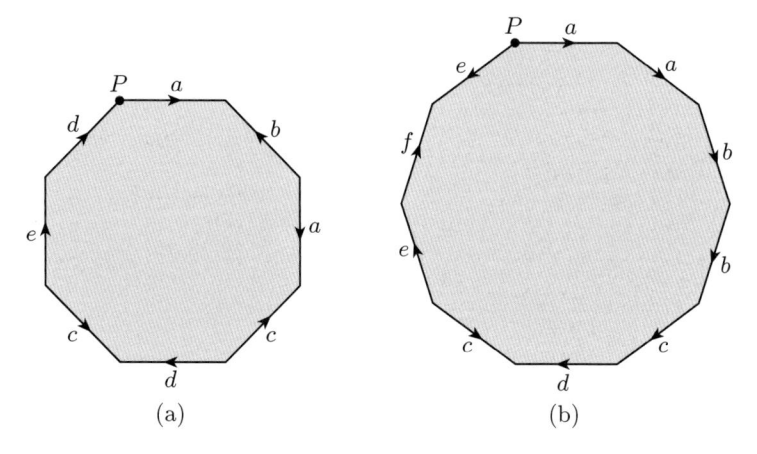

Figure 3.5

Problem 3.2

Draw the polygon with edge identifications that corresponds to the edge expression $c^{-1}a^{-1}ed^{-1}cbab$.

There are several arbitrary choices to be made when we derive an edge expression for a polygon with edge identifications: different choices lead to different edge expressions for the same polygon, and we shall treat such expressions as *equivalent*.

The starting edge can be chosen arbitrarily.
The effect of choosing a different starting edge is to permute the entries in the edge expression cyclically. For example, if we choose to start with the edge labelled c on the right of the polygon in Figure 3.1, we obtain the expression $c^{-1}a^{-1}ed^{-1}cbab$ instead of $abc^{-1}a^{-1}ed^{-1}cb$. Any two edge expressions related in this way are equivalent.

We call this operation *cycling* the edge expression.

The choice of distinct labels is arbitrary.
We could relabel the edges marked a in Figure 3.1 with a new letter x: the expression would then become $xbc^{-1}x^{-1}ed^{-1}cb$. Alternatively, we could relabel the edges marked a with b, and those marked b with a: the expression would then become $bac^{-1}b^{-1}ed^{-1}ca$. Any two expressions that can be made identical by relabelling edges are equivalent.

We call this operation *relabelling* the edges.

The choice of direction of the arrow on each edge is arbitrary, provided that the arrows on identified edges correctly indicate their relative senses.
If an edge appears twice, then we can reverse the arrow on one appearance provided that we also reverse the arrow on its other appearance. If an edge appears only once, we can reverse its arrow. The effect on the edge expression of reversing an arrow is to replace the corresponding edge label by its inverse, using the rule $(a^{-1})^{-1} = a$ whenever necessary. For example, if we change the directions of the arrows on both of the edges marked c in Figure 3.1, then we obtain the edge expression $abca^{-1}ed^{-1}c^{-1}b$, and this is equivalent to $abc^{-1}a^{-1}ed^{-1}cb$. If we now change the direction of the arrow on the only edge marked d, then we obtain another equivalent expression: $abc^{-1}a^{-1}edcb$.

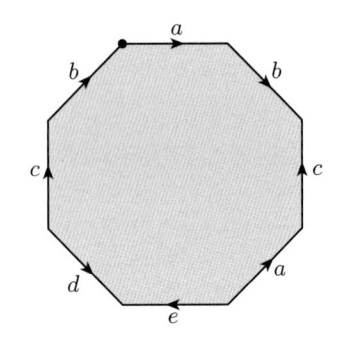

Figure 3.1 again

If the edge is not repeated, then the direction of the arrow is completely arbitrary.

The choice of direction in which to go round the edges of the polygon is arbitrary.
If we choose the anticlockwise direction for the surface in Figure 3.1, starting at the same vertex, we obtain the equivalent edge expression $b^{-1}c^{-1}de^{-1}acb^{-1}a^{-1}$.

Problem 3.3

In the polygon in Figure 3.1, suppose that the edge label a is replaced by b, the label b is replaced by c, and so on cyclically, ending with the replacement of the label e by a. Write down the new edge expression.

Recall that, if a polygon with edge identifications is to represent a surface, then each edge label can occur at most twice. Similarly, if an edge expression is to represent a surface, then each letter can occur at most twice (with or without the exponent -1). Provided that no edge appears more than twice, an edge expression tells us everything that the corresponding polygon tells us about the surface they both define. The edge expression enables us to draw a polygon with edge identifications that describes the surface completely, and we can then compute its characteristic numbers using the methods of Section 2. However, this information can also be gleaned directly from the edge expression itself, as we now see.

3.2 The characteristic numbers

In this subsection we shall see how to determine the characteristic numbers of a surface from an edge expression corresponding to the surface. We begin with the orientability number.

In Section 2 we showed that a surface defined by a polygon with edge identifications is orientable if and only if each repeated edge appears in both senses. We can restate this in terms of edge expressions by saying that a surface defined by a polygon with edge identifications is orientable if and only if each repeated edge x appears once with the exponent -1 and once without: that is, as $\cdots x \cdots x^{-1} \cdots$ or as $x^{-1} \cdots x \cdots$.

So the edge expression $abc^{-1}a^{-1}ed^{-1}cb$ tells us that the surface defined by Figure 3.1 is non-orientable, because the edge b occurs twice in the same sense. In general, when some symbol x appears as $\cdots x \cdots x \cdots$ or as $\cdots x^{-1} \cdots x^{-1} \cdots$, the corresponding surface is non-orientable ($\omega = 1$); otherwise the surface is orientable ($\omega = 0$).

From the edge expression, we can also obtain the number of vertices, edges and faces in the subdivision of the surface represented by the edge expression. We can thus calculate the Euler characteristic of the surface. The method for determining the number of vertices also allows us to calculate the boundary number of the surface. The following worked problem illustrates the technique.

Worked problem 3.1

Find the characteristic numbers χ, β and ω for the surface corresponding to the edge expression $abc^{-1}a^{-1}ed^{-1}cb$.

Solution

To find the Euler characteristic, we need to determine the number of vertices, edges and faces in the subdivision represented by the edge expression. We begin by determining the number of vertices, which we achieve in several steps.

1. Suppose that the starting vertex of the edge a is P. We insert the letter P in front of the letter a:

    $\underline{Pa}\,bc^{-1}a^{-1}ed^{-1}cb$.

    We can also insert the vertex P at the start of the edge a on its second appearance, as a^{-1}, in the edge expression. Since the edge a starts at P, a^{-1} must end at P, and so we write the letter P after a^{-1}:

    $Pabc^{-1}\underline{a^{-1}P}\,ed^{-1}cb$.

2. The vertex P is also the endpoint of the final edge, labelled b. Thus, we write P after b at the end of the edge expression:

    $Pabc^{-1}Ped^{-1}c\,\underline{bP}$.

    We can also insert the letter P at the end of the edge b on its other appearance, giving

    $Pa\,\underline{bP}\,c^{-1}a^{-1}Ped^{-1}cbP$.

The process of determining the number of vertices is essentially the method of inserting vertices from Section 2.

At each step of the process, we underline the letters involved.

$P \bullet \!\!\xrightarrow{a}\!\!$

Figure 3.6

$\xrightarrow{b}\!\! \bullet P$

Figure 3.7

3. Since we have now inserted P before c^{-1} in the expression, we can deduce that the edge c ends at P, so we insert P after the other appearance of c (see Figure 3.8):

$$PabPc^{-1}a^{-1}Ped^{-1}\underline{cP}bP.$$

Continuing in this way, we obtain (see Figure 3.8):

$Pa\underline{Pb}Pc^{-1}a^{-1}Ped^{-1}cPbP$ b starts at P;

$PaPbPc^{-1}\underline{Pa^{-1}}Ped^{-1}cPbP$ a ends at P;

$PaPbPc^{-1}Pa^{-1}Ped^{-1}\underline{Pc}PbP$ c starts at P.

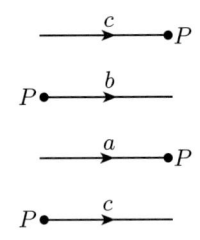

Figure 3.8

4. We have now inserted P into the expression at all possible places, leaving just one place left to insert a vertex, between e and d^{-1}. This must be a different vertex, which we label Q:

$$PaPbPc^{-1}Pa^{-1}P\underline{eQd^{-1}}PcPbP.$$

> Earlier, we called this new vertex U.

It follows that the subdivision of the surface represented by the edge expression has just two vertices, P and Q.

We can now compute the characteristic numbers χ, β and ω for the surface.

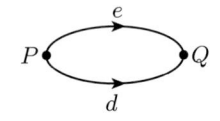

Figure 3.9

Euler characteristic χ

We count the numbers of vertices, edges and faces:
- *vertices:* there are 2 vertices (P, Q), so $V = 2$.
- *edges:* there are 5 edges (a, b, c, d, e), so $E = 5$.
- *faces:* there is only one edge expression, so $F = 1$.

So the Euler characteristic is $\chi = V - E + F = 2 - 5 + 1 = -2$.

> A single edge expression relates to a single polygon with edge identifications, and we have seen that this has just one face.

Boundary number β

The edges on the boundary are those that appear just once in a polygon with edge identifications, and these edges therefore appear just once in an edge expression. The edges d and e appear just once in the given edge expression, and so must lie on the boundary. The vertex insertion process tells us that this boundary component is represented by $PeQd^{-1}P$ (see Figure 3.9). Thus, they form a single boundary component, so $\beta = 1$.

Orientability number ω

We have already seen that the surface is non-orientable, because it contains a pair of edges (those labelled b) appearing in the same sense, as $\cdots b \cdots b \cdots$. Thus, $\omega = 1$. ∎

Remark

> In the process of inserting vertices, you need to be systematic. You should start by inserting a vertex at the left end of the edge expression. You need to insert all occurrences of a particular vertex before going on to the next. You should then move from left to right across the resulting expression to locate the first pair of edges between which no vertex has been inserted, and make this the starting point for inserting the next vertex.

Problem 3.4

For each of the following edge expressions, find the characteristic numbers of the corresponding surface.

(a) $abca^{-1}d^{-1}dg^{-1}cgf$ (b) $ahgkg^{-1}a^{-1}d^{-1}dk^{-1}f$

3.3 Edge equations

Our aim is to convert operations on polygons with edge identifications into algebraic form. It proves helpful in this regard to make edge expressions into equations. To see what we mean, consider the triangle in Figure 3.10, which has edge expression abc.

We can travel from the point P to the point Q by going directly along the edge labelled a, or by going (via R) along the edges labelled c and b against the indicated directions. We express this by the equation

$$a = c^{-1}b^{-1}.$$

We can rewrite this equation as

$$a = (bc)^{-1}$$

or, on multiplying both sides of the equation by bc, as

$$abc = 1.$$

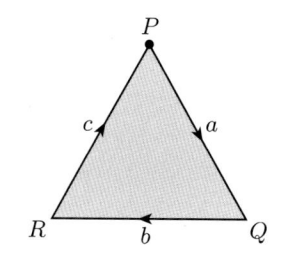

Figure 3.10

Recall that, algebraically, $(bc)^{-1} = c^{-1}b^{-1}$.

The left-hand side is the edge expression for the triangle in Figure 3.10. We can interpret this equation as saying that if we traverse the edges labelled a, b and c, then we return to our starting point, which is consistent with the triangle in Figure 3.10.

We call such an equation, in which an edge expression is equated to 1, an **edge equation**. For example, the edge expression

$$abc^{-1}a^{-1}ed^{-1}cd$$

has the corresponding edge equation

$$abc^{-1}a^{-1}ed^{-1}cd = 1.$$

This expresses the fact that, if we go around the polygon in Figure 3.11, starting from the marked vertex, then we return to the starting point.

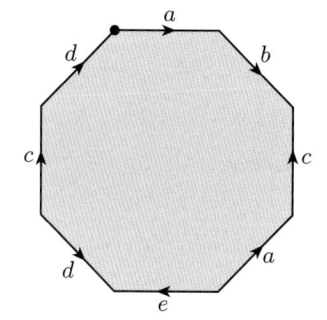

Figure 3.11

As we did in the case of the triangle in Figure 3.10, we can manipulate this equation algebraically, for example by bracketing the equation as

$$a(bc^{-1}a^{-1}ed^{-1}cd) = 1,$$

and so writing

$$a = (bc^{-1}a^{-1}ed^{-1}cd)^{-1} = d^{-1}c^{-1}(d^{-1})^{-1}e^{-1}(a^{-1})^{-1}(c^{-1})^{-1}b^{-1}$$
$$= d^{-1}c^{-1}de^{-1}acb^{-1}.$$

The inverse of a product is the product of the inverses *taken backwards*.

Interpreting this in terms of Figure 3.11, it tells us that going clockwise along the edge a at the top is the same as going anticlockwise round the other edges.

Problem 3.5

For the polygon with edge identifications in Figure 3.11, write down equations of the form $b = \cdots$ and $e = \cdots$.

Edge equations will prove very useful in the next unit, as we move towards proving the Classification Theorem for compact surfaces.

Surfaces constructed from several polygons

In *Unit B1*, we met the idea of gluing two compact surfaces together to create a new surface. We can use edge equations to represent this process.

Consider, for example, the two surfaces represented by polygons with edge identifications in Figure 3.12(a). Suppose we wish to glue the two surfaces together according to the identifications shown. One way would be to glue the polygons together along two edges labelled c, to create the single polygon with edge identifications in Figure 3.12(b).

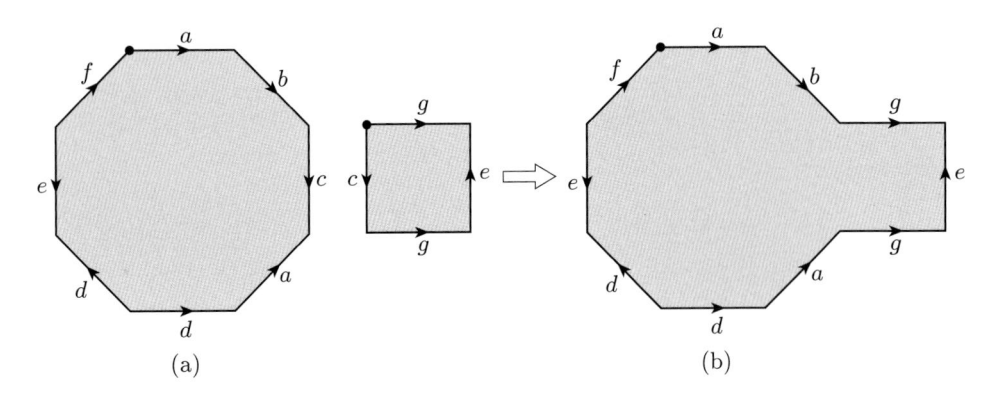

Figure 3.12

The original two polygons have edge equations
$$abca^{-1}d^{-1}de^{-1}f = 1 \quad \text{and} \quad ge^{-1}g^{-1}c^{-1} = 1.$$

This new polygon has edge equation
$$abge^{-1}g^{-1}a^{-1}d^{-1}de^{-1}f = 1.$$

We can obtain this third edge equation algebraically by rearranging the second edge equation $ge^{-1}g^{-1}c^{-1} = 1$ in the form $ge^{-1}g^{-1} = c$, and then substituting for c in the first edge equation to give
$$ab(ge^{-1}g^{-1})a^{-1}d^{-1}de^{-1}f = 1.$$

Thus the gluing process can be performed using edge equations by using one edge equation to obtain an expression for the edge along which the gluing is to take place and substituting this into the second edge equation.

Problem 3.6

The two surfaces represented by polygons with edge identifications in Figure 3.12(a) are to be glued together along the edges labelled e.

(a) Use the edge equations
$$abca^{-1}d^{-1}de^{-1}f = 1 \quad \text{and} \quad ge^{-1}g^{-1}c^{-1} = 1$$
for the two surfaces to obtain an edge equation for the glued surface.

(b) Draw a (single) polygon with edge identifications that represents the glued surface.

Some special edge equations

We conclude this section by collecting together the edge equations of some important surfaces. These are the surfaces obtained by the paper-and-glue constructions of *Unit B1*.

Unit B1, Subsection 1.3.

surface	polygon with edge identification	edge equation
cylinder		$aba^{-1}c = 1$
Möbius band		$abac = 1$
torus		$aba^{-1}b^{-1} = 1$
Klein bottle		$aba^{-1}b = 1$
projective plane		$abab = 1$
torus with 1 hole		$aba^{-1}cb^{-1} = 1$
2-fold torus		$aba^{-1}b^{-1}cdc^{-1}d^{-1} = 1$
sphere		$aa^{-1} = 1$

If we write $x = ab$ in the edge equation of a projective plane, we obtain the simpler equation $xx = 1$.

Notice that when we proceed from a torus (a sphere with one handle) to a 2-fold torus (a sphere with two handles) we add the extra letters $cdc^{-1}d^{-1}$. In general, when we add an expression of the form $xyx^{-1}y^{-1}$ to an edge equation, the effect is to add a handle to the corresponding surface.

More generally, if we add an edge expression of any of the forms given in the table to an edge equation of a surface, the result is to add a surface of the type corresponding to the edge expression to the original surface. In particular, we find that:

- when we add an expression of the form $xyx^{-1}y^{-1}$ to an edge equation, the effect is to add a handle to the surface;
- when we add an expression of the form xx to an edge equation, the effect is to add a cross-cap (or projective plane) to the surface;
- when we add an expression of the form xcx^{-1} to an edge equation, the effect is to pierce a hole (with boundary c) in the surface.

We shall need these results in *Unit B3*.

4 Strict subdivisions

After working through this section, you should be able to:
▶ explain the terms *strict subdivision* and *triangulation*;
▶ explain why two triangulations of a given compact surface have the same Euler characteristic.

In this section we return to general subdivisions of a compact surface, with or without boundary. In *Unit B1* we claimed that the Euler characteristic is a property of the surface, rather than of any particular subdivision — any two subdivisions of the surface have the same Euler characteristic. In this section we outline a proof of this. We first show that we can transform certain subdivisions into ones in which all the faces are triangles, called *triangulations*, and then that the original subdivision and the triangulation have the same Euler characteristic. We then sketch a proof that any two triangulations of the same compact surface have the same Euler characteristic, which thus gives the desired result.

If you are short of time, you may prefer to skim through this section on a first reading.

Unit B1, Theorem 3.1.

4.1 Triangulations

Throughout this subsection we shall consider compact surfaces, with or without boundary, represented as polygons with edge identifications.

Recall the definition of a subdivision.

Definition

A **subdivision** of a surface consists of a finite set of **vertices** and a finite set of **edges** such that:
- each vertex is an endpoint of at least one edge;
- each endpoint of an edge is a vertex;
- the vertices and edges form a connected graph;
- no two edges have any points in common;
- if the surface has a boundary, the boundary consists only of vertices and edges;
- the space obtained from the surface by removing the vertices and edges is a union of a finite number of disjoint pieces, called **faces**, each of which is homeomorphic to an open disc.

This definition appeared at the start of Section 1 of this unit and, originally, in Subsection 3.2 of *Unit B1*.

Recall also, from *Unit B1*, that:
- a vertex *belongs to* an edge if every open set containing the vertex contains points of the edge;
- a vertex *belongs to* a face if every open set containing the vertex contains points of the face;
- an edge *belongs to* a face if every open set containing the edge contains points of the face;
- two edges or faces *meet* at a vertex if there is a vertex that belongs to both of them;
- two faces *meet* at an edge if there is an edge that belongs to both of them;
- the edges that meet at a vertex are the *incident edges* of that vertex;
- a face *has* those vertices and edges that belong to it;
- if n edges and n vertices belong to a face, then the face *has* n edges and n vertices.

We refer to a face with n edges and n vertices as an *n-gon*, though when $n = 2, 3, \ldots 8$, we may prefer to use the terms *digon, triangle, quadrilateral, pentagon, hexagon, heptagon* and *octagon*, respectively.

Using this terminology, we can define a strict subdivision as follows:

Definition

A **strict subdivision** of a surface is a subdivision in which:
- any two faces meet at a single edge (together with the vertices that belong to this edge), at a single vertex, or not at all;
- each non-boundary edge belongs to two faces;
- each edge that is part of the boundary belongs to just one face;
- no face meets itself, either at a vertex or at an edge;
- the union of all the faces meeting a given vertex, together with the vertex and its incident edges, is homeomorphic to an open disc — or, if the vertex lies on the boundary, to an open half-disc.

Remarks

(i) These extra conditions that specify a subdivision as strict are needed for the formal proof of the equivalence of Euler characteristics for different subdivisions of the same compact surface. Since we provide only an outline proof, only the first two conditions will play any significant part in the discussion that follows.

(ii) The extra conditions rule out 1-gons (faces with one edge and one vertex) in a strict subdivision, though they may include n-gons for all $n \geq 2$.

Figure 4.1(a) shows a strict subdivision of a surface represented as a polygon with edge identifications. The two darker regions illustrate the last strict subdivision condition, showing that a union of faces meeting an interior vertex A is homeomorphic to an open disc and that a union of faces meeting a boundary vertex B is homeomorphic to an open half-disc.

Note that, from the definition, each union includes not only the faces but also the given vertex and its incident edges.

Figure 4.1(b) shows a subdivision of the same surface that is not strict, since the face F meets itself at two edges, the faces F and G meet at several edges, as do the faces F and H, and the lighter region is a union of faces meeting the boundary vertex C which is not homeomorphic to an open half-disc.

Of course, any *one* reason is sufficient to show that the subdivision in (b) is not strict.

(a) strict subdivision

(b) not a strict subdivision

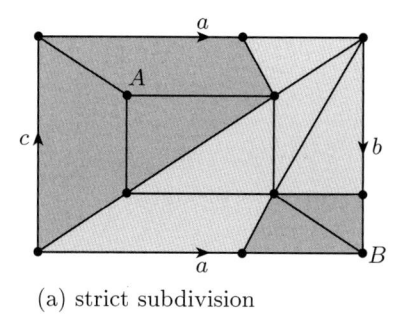

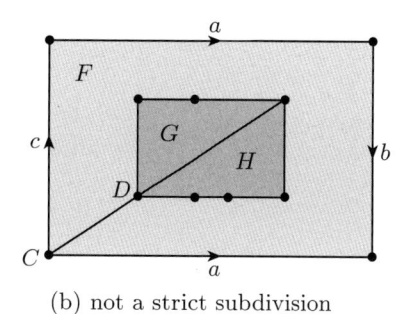

Figure 4.1

Particular care needs to be taken when checking subdivisions for strictness when the surface is represented as a polygon with edge identifications. For example, in Figure 4.1(b), face F meets itself not only at the edge CD but also at the edge labelled a because of the identification indicated. Another example is provided by Figure 4.2, which may superficially appear to be a strict subdivision. However, because of the identifications, faces F and G meet at an edge and also at a vertex not belonging to that edge, as do faces G and H, while faces F and H meet at three vertices but not at the edges joining them.

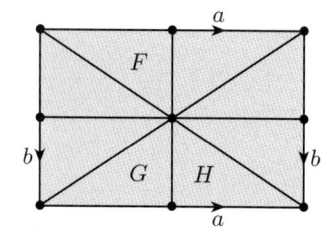

Figure 4.2

Problem 4.1

Determine whether the subdivisions in Figure 4.3 are strict.

(a)

(b)

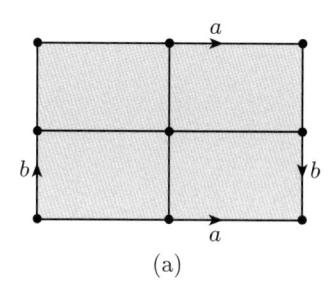

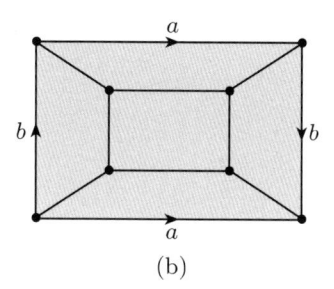

Figure 4.3

Of particular importance are those strict subdivisions that are made up entirely from triangles.

Definition

A **triangulation** is a strict subdivision in which each face is a triangle.

An example of a triangulation is shown in Figure 4.4.

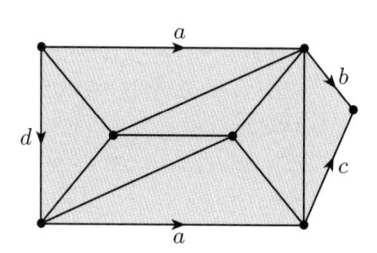

Figure 4.4

Our next task is to show how to obtain a triangulation from a given subdivision, by adding extra vertices and edges in a systematic manner. The first step is to see how we can triangulate an n-gon.

- For $n > 3$, we simply insert a new vertex into the interior of the n-gon and join it by edges to each of the original n vertices, as we illustrate in the case of a pentagon in Figure 4.5.
- For $n = 3$, we already have a triangle.
- For $n = 2$, we cannot proceed as for $n \geq 3$ since this would create two triangles but would violate the conditions for a strict subdivision, since the two triangles would meet at two edges, as Figure 4.6 illustrates. Instead, we add a vertex to each edge of the digon and join the two new vertices by an edge, as Figure 4.7 illustrates.

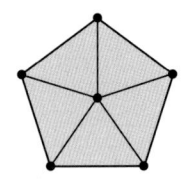

Figure 4.5

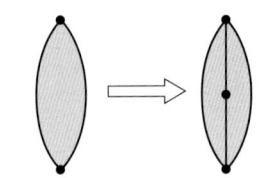

Figure 4.6

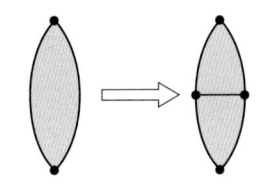

Figure 4.7

Provided there are no difficulties relating to edge identification, this procedure works very well for subdivisions in which there are no digons, as Figure 4.8 illustrates.

strict subdivision   triangulation

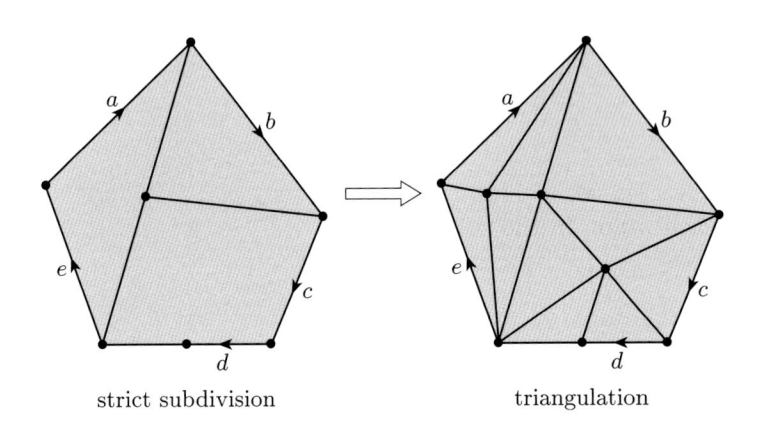

Figure 4.8

Problem 4.2

Convert the strict subdivision in Figure 4.9 into a triangulation.

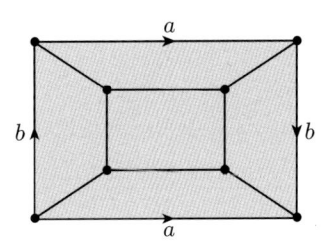

Figure 4.9

The key to the procedure working well for subdivisions in which there are no digons is that, for $n \geq 3$, the procedure makes changes that affect only the n-gon under consideration. However, in the case of a digon, the adding of two new vertices on the edges has an effect on the faces that meet the digon at those edges. For example, Figure 4.10 shows the effect of applying the procedure to a strict subdivision containing a digon. You will see that the resulting subdivision is not a triangulation, as it contains two quadrilaterals (faces F and G).

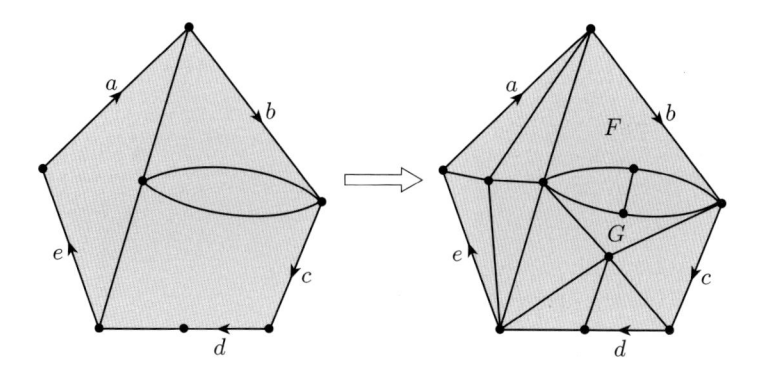

Figure 4.10

The simple way round this is to deal with all digons first, and then any n-gon converted into an $(n+1)$-gon by the triangulation of the digon will be triangulated as an $(n+1)$-gon and not an n-gon.

Finally, we need to consider the difficulties that can be caused by the edge identifications. You will recall, for example from Figure 4.2, that edge identifications can cause faces to meet in ways that violate the conditions for a strict subdivision: they can cause faces to meet at two edges, at an edge and a vertex not belonging to that edge, or at two vertices but not the edge joining them. Fortunately there is a simple technique for avoiding all these possibilities. First, triangulate each n-gon in the subdivision following the procedure described above (starting with the digons) to create a subdivision consisting entirely of triangles. Now place a vertex in the middle of each edge of each triangle, to create a subdivision consisting entirely of hexagons. Lastly, triangulate each hexagon. The result turns out to be a strict subdivision consisting entirely of triangles, that is, a triangulation. Figure 4.11 illustrates this final step applied to the non-strict subdivision consisting of triangles in Figure 4.2.

Dividing each edge in this way keeps the 'troublesome' vertices apart.

non-strict subdivision of triangles

triangulation

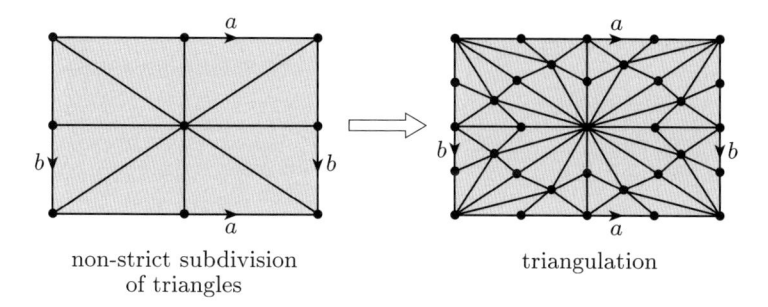

Figure 4.11

Of course, transforming each triangle into six smaller triangles in this way may not be the most efficient way of creating a triangulation — there may be many possible triangulations with far fewer faces — but it has the great advantage that it is systematic and always works.

Now we come to the pay-off, namely that, *by applying our systematic procedure to any subdivision of a surface, the triangulation we obtain has the same Euler characteristic as the original subdivision.* To demonstrate this we first note that, when we initially triangulate the subdivision, we deal with just one face at a time; so we analyse the effect on the Euler characteristic of dealing with each type of face. Suppose that the original subdivision has V vertices, E edges and F faces. When we initially triangulate an n-gon, three cases can arise:

Recall that the Euler characteristic is $\chi = V - E + F$.

- when the face is a digon ($n = 2$), we add 2 vertices, 3 edges and 1 face, so that the value of $V - E + F$ is unchanged;
- when the face is a triangle, we leave it unchanged, so the value of $V - E + F$ is unchanged;
- when the face is an n-gon for $n > 3$, we add 1 vertex, n edges and $n - 1$ faces (since 1 face is replaced by n faces), so the value of $V - E + F$ is unchanged.

Therefore, the new subdivision consisting entirely of triangles, created by triangulating each n-gon, has the same Euler characteristic as the original subdivision.

Suppose that the new subdivision has V^* vertices, E^* edges and F^* faces, where we know that $\chi = V - E + F = V^* - E^* + F^*$. Let us now consider the effect of converting each triangle in the subdivision into six smaller ones — the final step in our procedure. We first note that this final step has the effect of adding a vertex to each edge, a vertex to each face, and six edges to each face. Let us count the numbers of vertices, edges and faces in the final subdivision.

- *Vertices* The final subdivision has $V^* + E^* + F^*$ vertices, namely the V^* original vertices in the new subdivision, the E^* additional vertices on the edges, and the F^* additional vertices inside the faces.
- *Edges* Each edge of the new subdivision is divided into two parts and there are 6 additional edges in each face — thus, the final subdivision has $2E^* + 6F^*$ edges.
- *Faces* Each face is divided into 6 new faces — so the final subdivision has $6F^*$ faces.

Thus, the Euler characteristic of the final subdivision is

$$\begin{aligned}\chi &= \text{(number of vertices)} - \text{(number of edges)} + \text{(number of faces)} \\ &= (V^* + E^* + F^*) - (2E^* + 6F^*) + 6F^* \\ &= V^* - E^* + F^* = V - E + F.\end{aligned}$$

We have thus demonstrated that the triangulation obtained by applying our procedure to any subdivision of a compact surface has the same Euler characteristic as the original subdivision.

Problem 4.3

Verify that the triangulation you obtained in Problem 4.2 has the same Euler characteristic as the original subdivision.

4.2 Invariance of the Euler characteristic

We now explain why any two different triangulations of the same compact surface have the same Euler characteristic. This result is true in general, but we demonstrate it only in the case when the non-boundary edges of the first triangulation have only finitely many points in common with those of the second triangulation; this is called the *finite intersection property* for edges. To illustrate this property, consider Figure 4.13(a), which shows a part of each of two triangulations T_1 and T_2 of the same compact surface. If we superimpose T_2 on T_1 as in Figure 4.13(b) we can see that they have only finitely many points in common. For example, Figure 4.12 shows two triangles, one from each triangulation: the edges of the two triangles have just six points (A, B, C, D, E, F) in common, so the two triangles have the finite intersection property.

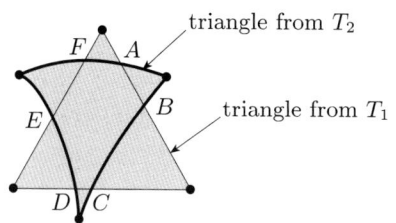

Figure 4.12

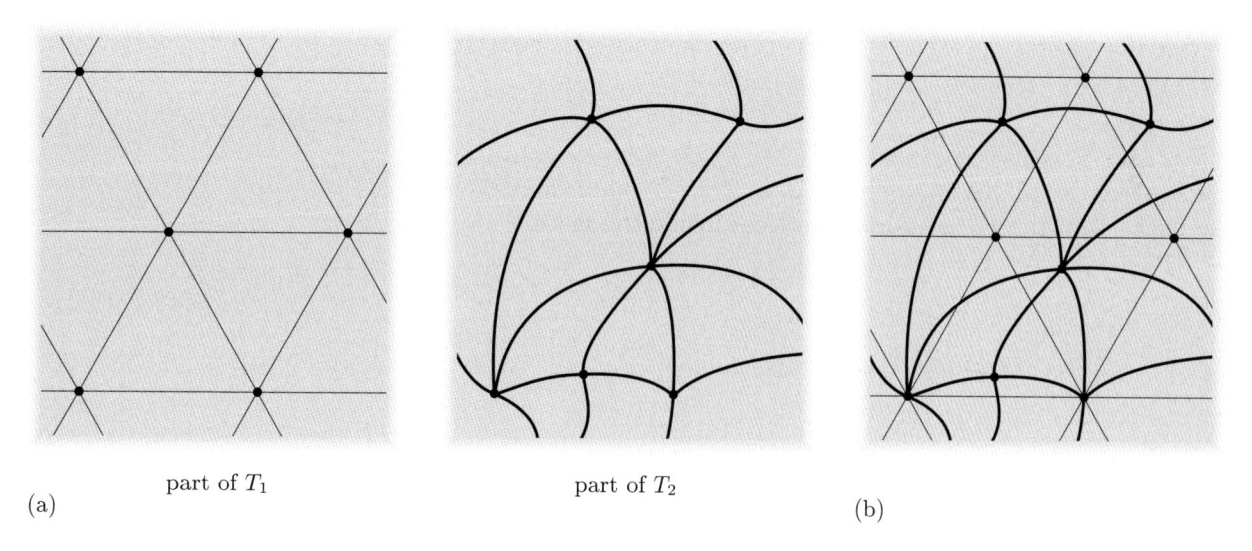

(a) part of T_1 part of T_2 (b)

Figure 4.13

We can combine any two such triangulations T_1 and T_2 to form a new subdivision $T_1 \cdot T_2$, as follows.

- *Vertices* The vertices of $T_1 \cdot T_2$ are the vertices of T_1, the vertices of T_2, and the points where the edges of T_1 intersect the edges of T_2: for example, in Figure 4.12 these consist of the vertices of the two triangles and the points A, B, C, D, E, F, as Figure 4.14 illustrates.
- *Edges* The edges of $T_1 \cdot T_2$ are the edges of T_1 and the edges of T_2, but divided up by the vertices of $T_1 \cdot T_2$: for example, the left-hand edge of the triangle from T_1 in Figure 4.12 is divided by the vertices E and F into three edges of $T_1 \cdot T_2$, as Figure 4.14 illustrates.
- *Faces* The faces of $T_1 \cdot T_2$ are the intersections of the faces of T_1 with those of T_2: for example, the face of the triangle from T_1 in Figure 4.12 is divided into four faces of $T_1 \cdot T_2$ by the triangle from T_2, as Figure 4.14 illustrates.

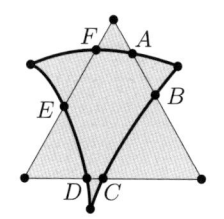

Figure 4.14

Figure 4.15 shows part of $T_1 \cdot T_2$ corresponding to the parts of T_1 and T_2 in Figure 4.13.

We now show that $T_1 \cdot T_2$ has the same Euler characteristic as T_1. We do so by adding edges to T_1 one at a time so as to produce $T_1 \cdot T_2$.

Suppose we have added a number of edges of $T_1 \cdot T_2$ to T_1 to obtain a subdivision S of the surface. Select any face f of S that is crossed by a curve that is a union of edges and vertices of $T_1 \cdot T_2$, such as the curves a, b and c in Figure 4.16.

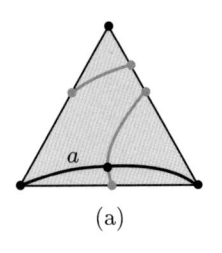

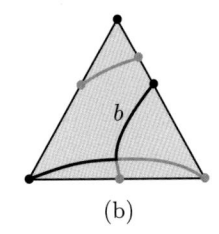

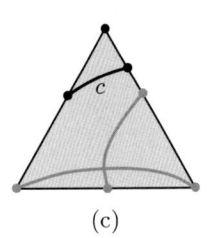

Figure 4.16

part of $T_1 \cdot T_2$

Figure 4.15

Add that curve, regarded as a single edge e. Then e meets the boundary of f in two points that may:

(a) both be vertices of f (e.g. Figure 4.16(a));
(b) be a vertex of f and a point on an edge of f (e.g. Figure 4.16(b));
(c) both be points on an edge or edges of f (e.g. Figure 4.16(c)).

Then adding the new edge and such new vertices as are necessary, and dividing f in two, we obtain a new subdivision S'. In each of the cases above, we have:

(a) 1 extra edge and 1 extra face (e.g. Figure 4.17(a));

(b) 1 extra vertex, 2 extra edges and 1 extra face (e.g. Figure 4.17(b));

(c) 2 extra vertices, 3 extra edges and 1 extra face (e.g. Figure 4.17(c)).

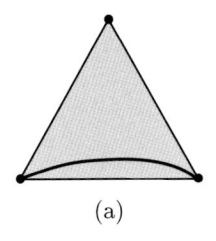

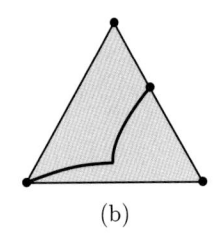

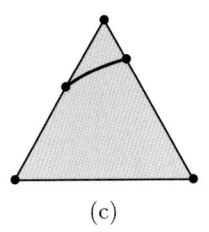

Figure 4.17

In each case, the change in $V - E + F$ is 0, so $\chi(S') = \chi(S)$.

By starting with T_1 and repeating this process a *finite* number of times, we eventually obtain $T_1 \cdot T_2$; thus,

This is where we need the finite intersection property.

$$\chi(T_1) = \chi(T_1 \cdot T_2).$$

Similarly, starting with T_2, we have

$$\chi(T_2) = \chi(T_1 \cdot T_2).$$

It follows that

$$\chi(T_1) = \chi(T_2),$$

as required.

We have shown that two triangulations of the same compact surface have the same Euler characteristic provided that the triangulations have the finite intersection property for edges. Although two triangulations of the same surface usually have this property, there are exceptions. Fortunately the proof can be extended to cover these exceptional cases, but it is difficult and so we omit it here. We conclude that *any two triangulations of the same compact surface have the same Euler characteristic.*

We can now deduce the promised result that any two subdivisions S_1 and S_2 of a given compact surface have the same Euler characteristic. We saw in Subsection 4.1 that any subdivision can be converted into a triangulation with the same Euler characteristic. Thus, given any two subdivisions S_1 and S_2 of the same surface, we can find triangulations T_1 and T_2 such that $\chi(S_1) = \chi(T_1)$ and $\chi(S_2) = \chi(T_2)$. But, as we have just shown, $\chi(T_1) = \chi(T_2)$. Thus, S_1 and S_2 have the same Euler characteristic.

The Euler characteristic therefore is a property of the surface, and not of the subdivision. We can calculate the Euler characteristic by using any convenient subdivision — in particular, we can use the subdivision corresponding to a representation of the surface as a polygon with edge identifications — and any such representation gives the same value.

Solutions to problems

1.1
tetrahedron: $\quad 2 = 6(\frac{2}{3} + \frac{2}{3} - 1) = \frac{1}{2}.4.3\,(\frac{2}{3} + \frac{2}{3} - 1)$
$\quad\quad\quad\quad\quad = \frac{1}{2}.4.3\,(\frac{2}{3} + \frac{2}{3} - 1);$

cube: $\quad 2 = 12(\frac{2}{3} + \frac{2}{4} - 1) = \frac{1}{2}.6.4\,(\frac{2}{3} + \frac{2}{4} - 1)$
$\quad\quad\quad\quad\quad = \frac{1}{2}8.3(\frac{2}{3} + \frac{2}{4} - 1);$

octahedron: $\quad 2 = 12(\frac{2}{4} + \frac{2}{3} - 1) = \frac{1}{2}.8.3\,(\frac{2}{4} + \frac{2}{3} - 1)$
$\quad\quad\quad\quad\quad = \frac{1}{2}.6.4\,(\frac{2}{4} + \frac{2}{3} - 1);$

dodecahedron: $\quad 2 = 30(\frac{2}{3} + \frac{2}{5} - 1) = \frac{1}{2}.12.5(\frac{2}{3} + \frac{2}{5} - 1)$
$\quad\quad\quad\quad\quad = \frac{1}{2}.20.3\,(\frac{2}{3} + \frac{2}{5} - 1);$

icosahedron: $\quad 2 = 30(\frac{2}{5} + \frac{2}{3} - 1) = \frac{1}{2}.20.3\,(\frac{2}{5} + \frac{2}{3} - 1)$
$\quad\quad\quad\quad\quad = \frac{1}{2}.12.5\,(\frac{2}{5} + \frac{2}{3} - 1).$

1.2
Multiplying the first regular subdivision formula by jk, we obtain
$$\chi jk = E(2k + 2j - jk),$$
from which the formula for E follows.
The other two results are obtained similarly.

1.3
It follows from the diagram below that the tetrahedron is self-dual.

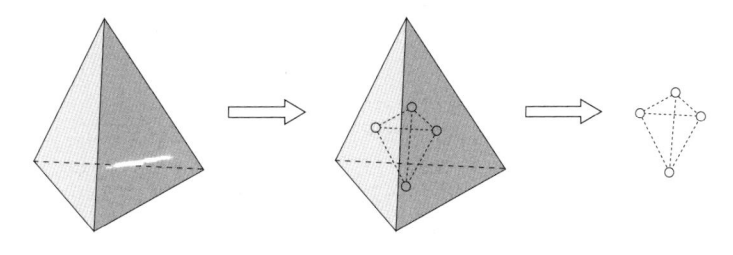

1.4 (a)

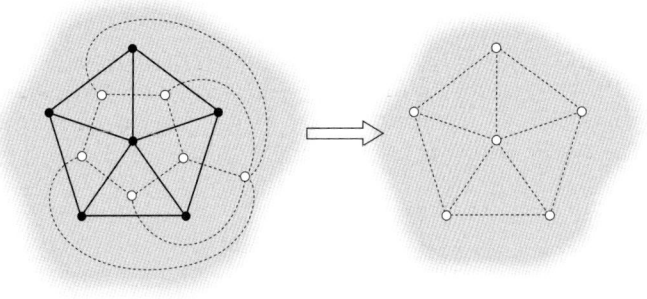

Notice how we can redraw the dual subdivision to show that it is self-dual.

(b)

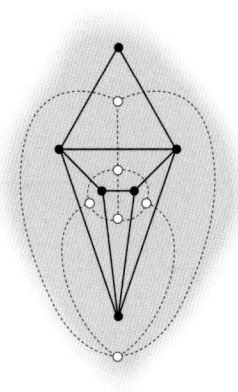

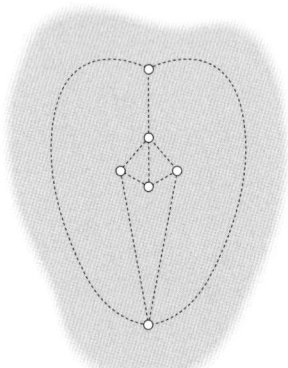

1.5

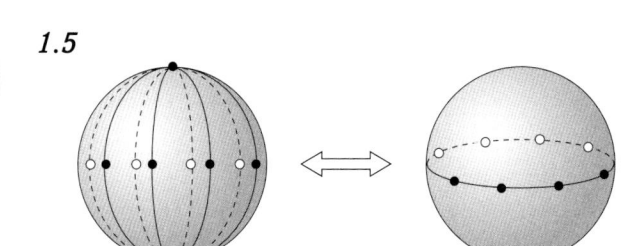

1.6

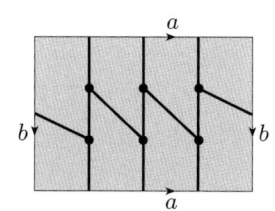

$(j,k) = (3,6)$
$V = 6, E = 9, F = 3$

$(j,k) = (4,4)$
$V = 3, E = 6, F = 3$

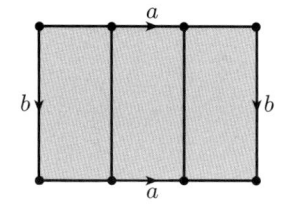

$(j,k) = (6,3)$
$V = 3, E = 9, F = 6$

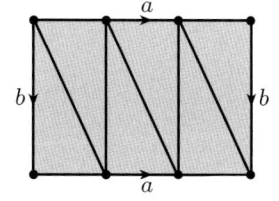

1.7 (a) Since such a surface has a regular subdivision with $(j,k) = (3,7)$, it also has one with $(j,k) = (7,3)$, by duality (Theorem 1.4).

(b) When $j = k = 5$ we have $jk - 2j - 2k = 5$, so that (by Corollary 1.2) V, E and F are all positive integers. So, by Theorem 1.3, every surface with negative Euler characteristic has a regular subdivision with $(j,k) = (5,5)$.

1.8 When $k = 5$ and $j = 4$, we have
$$jk - 2j - 2k = 20 - 8 - 10 = 2,$$
and the regular subdivision formulas give
$V = \frac{30}{2} = 15$, $E = \frac{60}{2} = 30$, $F = \frac{24}{2} = 12$.
Since these are all positive integers, there is such a regular subdivision by Theorem 1.3.

1.9 When $j = 4$, we have $2(1-\chi)j/(j-2) = 12$, and so $4 \leq k \leq 12$. Also,
$$V = \frac{2k}{k-4}, \quad E = \frac{4k}{k-4}, \quad F = \frac{8}{k-4}.$$
The value of F is an integer when $k - 4$ divides 8, so we can only have $k = 5, 6, 8, 12$. All four values give positive integer values for V, E and F. We thus obtain the following table of regular subdivisions with $j = 4$ and $k \geq j$.

j	k	V	E	F
4	5	10	20	8
4	6	6	12	4
4	8	4	8	2
4	12	3	6	1

Their duals correspond to the following values.

j	k	V	E	F
5	4	8	20	10
6	4	4	12	6
8	4	2	8	4
12	4	1	6	3

When $j = 5$, we have $2(1-\chi)j/(j-2) = 10$, and so $5 \leq k \leq 10$. Also,
$$V = \frac{4k}{3k-10}, \quad E = \frac{10k}{3k-10}, \quad F = \frac{20}{3k-10}.$$
The value of F is an integer when $3k - 10$ divides 20, so we can only have $k = 5, 10$. Both values give positive integer values for V, E and F. We thus obtain the following table of regular subdivisions with $j = 5$ and $k \geq j$.

j	k	V	E	F
5	5	4	10	4
5	10	2	5	1

The first of these is self-dual. For the second, we can deduce the existence of the dual subdivision, with $k = 5$ and $j \geq k$, as follows.

j	k	V	E	F
10	5	1	5	2

When $j = 6$, we have $2(1-\chi)j/(j-2) = 9$, and so $6 \leq k \leq 9$. Also,
$$V = \frac{k}{k-3}, \quad E = \frac{3k}{k-3}, \quad F = \frac{6}{k-3}.$$
The value of F is an integer when $k - 3$ divides 6, so we can only have $k = 6, 9$. The values of V and E are positive integers only when $k = 6$. We thus obtain the following regular subdivision, which is self-dual.

j	k	V	E	F
6	6	2	6	2

1.10 (a) To have $V = F$, the subdivision formulas (Corollary 1.2) tell us that we must have $j = k$ (so the subdivision must be self-dual). Putting $j = k$ and $F = 1$ into the formula
$$F = \frac{2(-\chi)j}{jk - 2j - 2k}$$
gives
$$2(-\chi)j = j^2 - 4j,$$
and hence
$$j = k = 4 - 2\chi.$$
But then the subdivision formula for E gives
$$E = \frac{-\chi j}{j-4} = \frac{j}{2} = 2 - \chi.$$
Since this is a positive integer, these values of j and k determine a regular subdivision with $2 - \chi$ edges, by Theorem 1.3.

(b) When $\chi = 2 - 2n$, the number of edges is
$$E = 2 - \chi = 2 - (2 - 2n) = 2n.$$

1.11 We have $k = 5$ and $\chi = -6$. We consider values of j with
$$3 \leq j \leq \frac{2(1-\chi)k}{k-2} = \frac{70}{3},$$
that is, with $3 \leq j \leq 23$.

The regular subdivision formulas become
$$V = \frac{60}{3j-10}, \quad E = \frac{30j}{3j-10}, \quad F = \frac{12j}{3j-10}.$$

From the formula for V, the possible values of j are 4, 5 and 10.

When $j = 4$, we have $V = 30$, $E = 60$, $F = 24$.
When $j = 5$, we have $V = 12$, $E = 30$, $F = 12$.
When $j = 10$, we have $V = 3$, $E = 15$, $F = 6$.

Since V, E and F are all positive integers in each case, these all give regular subdivisions.

2.1 There is only one face, so $F = 1$.

There are 6 distinct edges (a, b, c, d, e, f), so $E = 6$.

Labelling the vertices, we obtain the picture below, so $V = 3$.

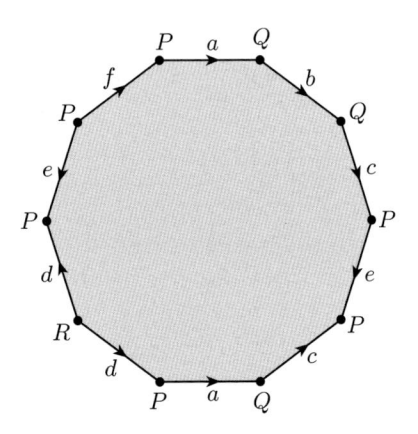

The Euler characteristic is therefore
$$\chi = V - E + F = 3 - 6 + 1 = -2.$$

2.2 **(a)** The surface has 1 face and 3 edges (a, b, c). We label the vertices as shown, so there are 2 vertices.

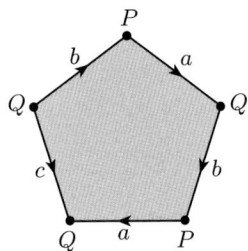

It follows that the Euler characteristic of the surface is $\chi = 2 - 3 + 1 = 0$.

(b) The surface has 1 face and 3 edges (a, b, c). We label the vertices as shown, so there is only 1 vertex.

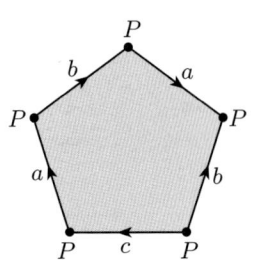

It follows that the Euler characteristic of the surface is $\chi = 1 - 3 + 1 = -1$.

(c) The surface has 1 face and 3 edges (a, b, c). We label the vertices as shown, so there are 3 vertices.

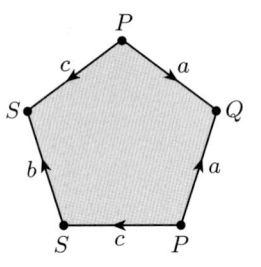

It follows that the Euler characteristic of the surface is $\chi = 3 - 3 + 1 = 1$.

2.3 The edges that appear only once are b (which is incident only to Q) and f (which is incident only to P). So these edges form two unconnected components of the boundary, and $\beta = 2$.

2.4 **(a)** The edge c is the only unrepeated edge, so $\beta = 1$.

(b) The edge c is the only unrepeated edge, so $\beta = 1$.

(c) The edge b is the only unrepeated edge, so $\beta = 1$.

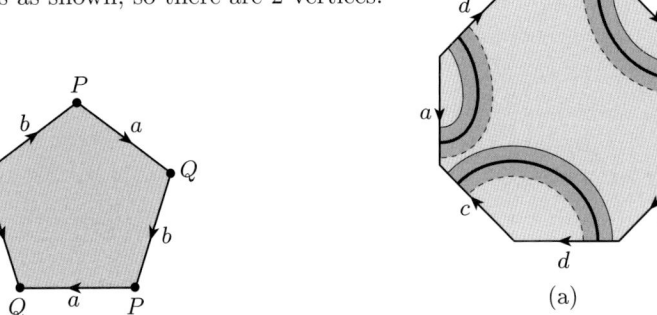

2.5

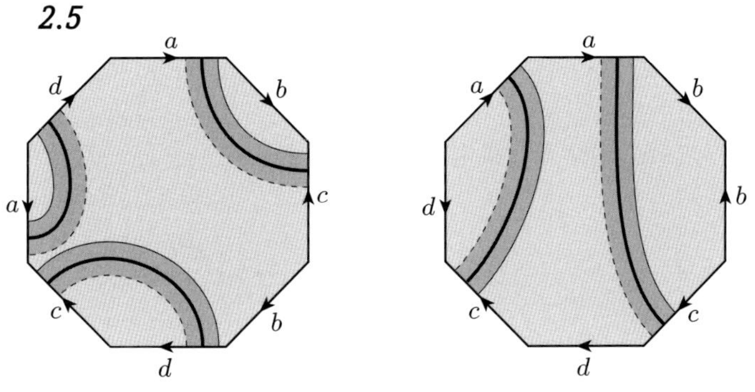

(a) The curve has a neighbourhood with one boundary component, which is therefore a Möbius band. The surface is not orientable.

(b) The curve has a neighbourhood with two boundary components, which is therefore a cylinder. From this evidence we cannot determine whether the surface is orientable. (In fact it is not, as you can see by drawing a thickened curve joining the two edges labelled a.)

2.6 **(a)** The edges labelled a occur in the same sense, and so there is a closed curve whose thickened neighbourhood is a Möbius band. Thus, the surface is non-orientable. (The edges labelled d can also be joined by a Möbius band.)

(b) The edges labelled d occur in the same sense, and so there is a closed curve whose thickened neighbourhood is a Möbius band. Thus, the surface is non-orientable.

2.7 We use Theorem 2.1.

(a) There are two edges occurring in the same sense (the edges labelled b, or the edges labelled c), so the surface is non-orientable,

(b) All pairs of identified edges occur in opposite senses, so the surface is orientable.

(c) There are two edges occurring in the same sense (the edges labelled b), so the surface is non-orientable,

2.8 **(a)** The edges labelled a (or those labelled b) occur in the same sense, so the surface is non-orientable and $\omega = 1$.

(b) The edges labelled a occur in the same sense, so the surface is non-orientable and $\omega = 1$.

(c) The edges in each identified pair (a and c) occur in opposite senses, so the surface is orientable and $\omega = 0$.

3.1 **(a)** $ac^{-1}bac^{-1}$

(b) $aa^{-1}bcd^{-1}b^{-1}$

3.2

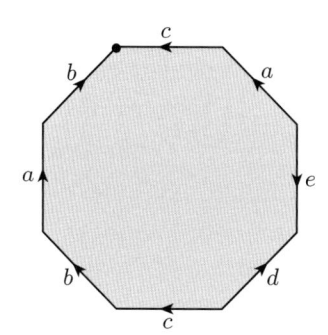

3.3 $bcd^{-1}b^{-1}ae^{-1}dc$

3.4 **(a)** From the edge expression $abca^{-1}d^{-1}dg^{-1}cgf$, we obtain

$\underline{Pa}\ bc\ \underline{a^{-1}P}\ d^{-1}dg^{-1}cg\ \underline{fP}$ (a starts at P)
$Pabca^{-1}Pd^{-1}\ \underline{dP}\ g^{-1}cgfP$ (d ends at P)
$Pabca^{-1}Pd^{-1}dPg^{-1}c\ \underline{gP}fP$ (g ends at P)
$P\ \underline{aQ}\ bc\ \underline{Qa^{-1}}\ Pd^{-1}dPg^{-1}cgPfP$ (a ends at Q)
$PaQbcQa^{-1}Pd^{-1}dPg^{-1}\ \underline{cQ}\ gPfP$ (c ends at Q)
$PaQbcQa^{-1}Pd^{-1}dP\ \underline{g^{-1}Q}\ cQgPfP$ (g starts at Q)
$PaQb\ \underline{Qc}\ Qa^{-1}Pd^{-1}dPcQgPfP$ (c starts at Q)
$PaQbQcQa^{-1}P\ \underline{d^{-1}R}\ dPg^{-1}QcQgPfP$ (d starts at R)

Euler characteristic χ

The surface thus has 3 vertices (P, Q, R), 6 edges (a, b, c, d, g, f) and 1 face.

It follows that the Euler characteristic of the surface is $\chi = 3 - 6 + 1 = -2$.

Boundary number β

The letters b and f appear just once, and there are two boundary components, QbQ and PfP; so $\beta = 2$.

Orientability number ω

The edge c occurs twice in the same sense, so the surface is non-orientable: $\omega = 1$.

(b) Inserting vertices, we obtain
$$PaQhQgPkPg^{-1}Qa^{-1}Pd^{-1}RdPk^{-1}PfP.$$

Euler characteristic χ

The surface has 3 vertices (P, Q, R), 6 edges (a, h, d, k, f, g), and 1 face.

So the Euler characteristic is $\chi = 3 - 6 + 1 = -2$.

Boundary number β

The letters h and f appear just once, and there are two boundary components, QhQ and PfP; so $\beta = 2$.

Orientability number ω

Each repeated letter appears once with the exponent -1, and once without, so the surface is orientable: $\omega = 0$.

3.5 $b = a^{-1}(c^{-1}a^{-1}ed^{-1}cd)^{-1}$
$= a^{-1}d^{-1}c^{-1}de^{-1}ac;$
$e = acb^{-1}a^{-1}d^{-1}c^{-1}d.$

3.6 **(a)** We rearrange the edge equation $ge^{-1}g^{-1}c^{-1} = 1$ first as
$$e^{-1}g^{-1}c^{-1} = g^{-1}$$
and then as
$$e^{-1} = g^{-1}(g^{-1}c^{-1})^{-1} = g^{-1}cg.$$

We substitute this into the other edge equation to obtain
$$abca^{-1}d^{-1}dg^{-1}cgf = 1.$$

(b)

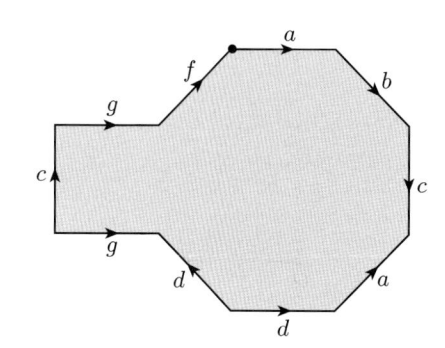

4.1 (a) This is not a strict subdivision since face F meets face I at an edge and at two vertices not belonging to that edge, and face G meets face H in the same manner.

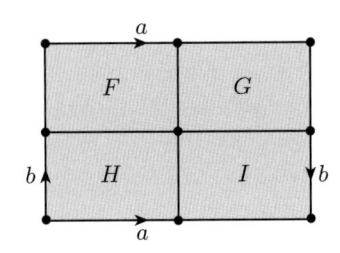

(b) This is a strict subdivision.

4.2

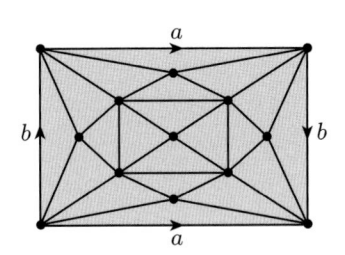

4.3 The original subdivision has 5 vertices, 10 edges and 5 faces, and the Euler characteristic is
$$\chi = 5 - 10 + 5 = 0.$$
The new subdivision has 10 vertices, 30 edges and 20 faces, and the Euler characteristic is
$$\chi = 10 - 30 + 20 = 0,$$
which is the same as before.

Index

boundary number 24

edge expression 31
Euler characteristic 23
   invariance 44

finite intersection property 44

inserting vertices 23, 34

method of inserting vertices 34

regular subdivision formulas 8

regular subdivisions
   sphere 12
   torus 15

sphere
   regular subdivisions 12
subdivision
   dual 11
   regular 7

torus
   regular subdivisions 15
triangulation 41